IFLA

International Federation of Library Associations and Institutions
Fédération Internationale des Associations de Bibliothécaires et des Bibliothèques
Internationaler Verband der bibliothekarischen Vereine und Institutionen
Международная Федерация Библиотечных Ассоциаций и Учреждений

IFLA Publications 18

Studies in the International Exchange of Publications

Edited by Peter Genzel

K·G·Saur München·New York·London·Paris 1981

IFLA Publications
edited by Willem R.H. Koops

Recommended catalog entry:

Studies in the international exchange of publications / ed. by Peter Genzel. – München; New York etc.: K.G. Saur, 1981. – 124 p.; 21 cm. – (IFLA Publications; 18)
ISBN 3-598-20379-9

Studies in the international exchange of publications /ed. by Peter Genzel. – München ; New York ; London ; Paris : Saur, 1981.
(IFLA publications ; 18)
ISBN 3-598-20379-9)

NE: Genzel, Peter [Hrsg.] ; International Federation of Library Associations: IFLA publications

ISSN 0344–6891 (IFLA publications)

Printed and bound in the Federal Republic of Germany
for K.G. Saur Verlag KG, München
by Friedrich Pustet, Regensburg

In memory of
ISTVÁN GOMBOCZ
Colleague and Friend

Contents

Foreword

It is not uncommon for the papers presented on different occasions by members of a scientific body to be published in a collective volume. In this way good ideas and important research work are prevented from falling into oblivion. At the same time suggestions for practical work can be given and the papers can also serve as a basis for further investigation.

The present papers were prepared for, and read at, the public meetings of the Section on the Exchange of Publications (since 1980: Section on Exchange and Acquisition) from 1977 to 1979, so they reflect the latest findings. Most of them have been revised, completed, and brought up to date for the purpose of this publication.

The Standing Committee of the Section assembled by its late chairman, István Gombocz, has remained fairly stable so that almost all the same people have co-operated for almost a decade. This fact favours continuity, specialization and variety. The wide scope of themes is the result.

I want to thank all my colleagues for the great trouble they have taken, and to Alex Allardyce I am specially obliged for clarifying the English of the texts where necessary.

Peter Genzel

Universal Availability of Publications and the Section on the Exchange of Publications. Report on its activities 1970–1979*

Peter Genzel**

1. Introduction

Since the first days of its existence (1930) the Section on the Exchange of Publications, then under a different name, has concentrated its activities on aspects of what we today term UAP. For many years the exchange of theses had been the main subject, but in course of time the field of action was extended to other material, such as official publications, bibliographies, commercial publications, etc., and also the organization of exchanges. In this connection the Convention concerning the International Exchange of Publications and the Convention concerning the Exchange of Official Publications and Government Documents between States (Unesco, 1958) are to be considered landmarks.

Today the activities are chiefly focussed on the examination of the main aspects of exchange practice and the search for the best working methods with a view to their generalization and to passing on many decades of experience in order to improve the exchange of publications on a world-wide scale. In this sense the entire present activities of the Section are more than ever a contribution to UAP. They are based on IFLA's Medium-term Programme (The Hague, 1976; paragraph on the "Exchange of publications", p. 24–25, and elsewhere) which is itself the foundation of the Section's own Programme.[1]

Since the first four decades (1930–1969) of the activities of the Section have already been dealt with in detail[2], this report has as its starting point the year 1970 and can be regarded as a kind of continuation.

2. The Projects

2.1. General, comprehensive

2.1.1. Theory and practice, standardization

One of the outstanding events during the period under review was the second *European Conference of exchange experts* which was organized by IFLA's then Committee on the Exchange of Publications, with Unesco assistance, at the Austrian National Library in Vienna in 1972.[3] 29 European experts and 14 non-European and Austrian observers attended the meeting at which 18 working papers were presented. The first group of themes was devoted to "Exchange and other means of access to world literature", six papers dealt with the need to increase exchanges with developing countries and a further group of topics was devoted to current exchange problems. The resolutions concentrated on the relationship of exchanges with loans, the implementation of Unesco's Exchange Conventions and the solution of several practical problems of immediate interest. Some important resolutions were included in the Section's Medium-term Programme and have since been implemented. The Conference will be referred to again throughout this paper.

At this conference, a paper of general interest was read by *Z. Jabłoński* on "The Role of international exchange in the library system".[4] The author and the audience came to the conclusion that international exchange is a form of co-operation, which could be done by a central organization, or it could be decentralized. The latter is in some ways simpler and has the advantage of facilitating direct contacts between similar institutes. Centralization guarantees co-ordination, but can degenerate into bureaucracy. Exchange and purchase form a unity, and in many libraries exchanges are equal to purchases in quantity. Therefore it is not sufficiant to try to co-ordinate the former without the latter. Although international exchange is often more expensive than purchase, many countries have to use exchange because of other conditions, such as e. g. limitations of foreign currency.

Two other significant papers were written by *B. P. Kanevsky.* In "The international exchange of publications and the free flow of books"[5] he says that by speeding up the development of international scientific, technical and cultural relations, the contemporary scientific and cultural revolution stimulates the free flow of publications through international exchange. He then surveys the present situation of international exchange not leaving out of consideration some difficulties still occasionally encountered. Suggestions for developing the international exchange of publications are outlined.

Some of these problems are resumed, developed and brought up-to-date in "The book world and the international book exchange".[6] After a short survey of the present trends of book production Kanevsky summarizes the results gathered from the answers to a questionnaire sent to libraries in 80 countries. The main obstacles lie in the social and economic spheres (e. g., special economic situation in the developing countries, continuous rise of book prices, etc.). They must be overcome as well as the somewhat neutral attitude of librarians towards the application of modern techniques in the exchange of publications (microforms etc.). Moreover, he pleads for a better utilization of existing book exchange stocks. We cannot but agree with his optimistic view on the future of exchanges, since the international situation favours the development of economic, scientific and cultural relations between countries with different social systems.

An important success is the publication by Unesco of the 4th edition of the *"Handbook on the international exchange of publications"* in 1978.[7] The first plan for the manuscript had been drafted by I. Gombocz on the basis of a contract between Unesco and IFLA. F. Vanwijngaerden is the editor, and most of the contributors are members of the Section's Standing Committee. The authors of the main chapters are P. Genzel, J. Lethève, D. Oertel, V. Popov and M. J. Schiltman. The work plan, the time schedule, the questionnaire, and the manuscripts were discussed at two special meetings convened on behalf of IFLA in the Federal Republic of Germany (1973) and the German Democratic Republic (1974). Compared with the former editions, the new one has been changed considerably in some parts and made more specific in others. The Handbook aims at providing a guide on the methodology, organization and management of the international exchange of publications by bringing up-to-date the text of the 3rd edition and by tracing the development of the last ten years, especially as regards standardization and the part played by international organizations in promoting the exchange of publications. Moreover, it is a detailed up-to-date directory of exchange centres with a national responsibility, providing practical information on their activities and services, which will enable libraries all over the world to establish contacts with appropriate exchange partners. The Handbook is intended for the use of: librarians who want to set up an exchange centre, especially in developing countries; exchange librarians who want to enlarge their knowledge; students at library school; officers responsible for library work; information officers and documentalists.

Whereas the background and practical handling of exchanges are described in this Handbook, actual implementation of the exchange of publications by written communication between partners will be more easily effected by a new letter-writer in five languages which was presented in 1978: *"Letters for the international exchange of publications"*.[8] In a way it can be regarded as a supplement to the Handbook. The basic English text was compiled by A. Allardyce, the translations were prepared by B. P. Kanevsky, J. Lethève, M. Razumovsky and A. Rodríguez, P. Genzel served as initiator and editor. The layout is as follows: there are five self-contained language parts with English as the basic text followed by the translated/adapted texts. Within each individual language part the arrangement of the subject-matter is the same throughout so that equivalents can easily be identified. The book is intended to enable the exchange librarian to maintain the elementary routine correspondence without other assistance.

It can also by regarded as a modest contribution to standardization in international librarianship. The "Project for standard book-exchange request forms" by *J. Lethève*[9] has the same objective. The author extended this project to some other forms frequently used in exchanges. His paper on the standardization of forms read in Vienna[10] gives an outline of the present situation, the barriers to be overcome, and the advantages of standardization. Moreover, forms for initiating exchanges, lists of offers of publications, requests for publications, and claims are presented. From the discussion it was clear that international standardization is difficult, but must keep pace with automation.

2.1.2. The Exchange Conventions of Unesco (1958)

One of the most far-reaching obligations untertaken by the Section is to follow the working of the Exchange Conventions. The European Conference of exchange experts was in fact devoted to this problem, particularly the "Inquiry into the execution and results of Unesco's Exchange Conventions fourteen years after their adoption" carried out by *I. Gombocz* on the basis of a questionnaire.[11] At first sight the results were rather discouraging. Although some important states with highly developed exchange practice had ratified the Conventions, others had abstained. But many of the latter practised the principles of the Conventions. It seemed, however, more serious that so many developing countries were abstaining in spite of the evident advantages which they might gain. Nevertheless, one could not state that the Conventions should be considered a failure. Some useful measures were recommended to Unesco.

Similarly, in his paper on "The multilateral Unesco Exchange Conventions (1958) and their international significance" *V. Popov*[12] presents the case for the Conventions. In his analysis of the negative or undecided attitude still assumed by many countries he comes to the conclusion that the Conventions have not yet been given sufficient publicity, that in some cases there is uncertainty and even ignorance as to the benefits of adherence and that the Conventions are occasionally underestimated. The author suggests how Unesco's General Information Programme may be supported and the UAP Programme promoted by adherence to, and realization of, the Conventions.

A resolution to this effect passed by the Section at the Brussels meeting resulted in a Contract signed between Unesco and IFLA on a *"Study on the practical effects of existing international conventions on the exchange of publications"* to be carried out by the Section on Exchanges. In the 1st part the existing conventions will be analyzed, with particular emphasis on those of 1958. The 2nd part will contain the main statistical data and other pertinent information on exchange to be gathered by means of a questionnaire. An overall analysis of the principle results of the application of the Conventions will be given and the progress made over the last few years be shown in the 3rd part.

Finally, mention must be made of a useful preliminary study by *V. Popov* for the second part of the above-mentioned Handbook on the international exchange of publications: "Short surveys of the European exchange centres. General information on organization and working methods of the individual exchange centres".[13] This theme must also be seen in the context of the Unesco Conventions of 1958 since they include articles concerning national exchange centres.

2.1.3. Bibliography of world literature on the international exchange of publications

The *"Bibliographie 1960 – 1970. Echanges internationaux de publications"* by B. P. Kanevsky and F. Vanwijngaerden[14] is the continuation of "Echanges internationaux de publications. Bibliographie. 1817–1960" compiled by J. L. Dargent in two volumes. The new bibliography comprises 450 entries from 24 countries and has a selective character. It reveals the sources relating to the history of the international book exchange, traces the development of international agreements and presents the relevant national legislation. Moreover, the bibliography gives an idea of the problems connected with exchanges and may serve as a stimulus for a more profound analysis and further development of the international exchange of publications. The Royal Library in Brussels has been charged with the task of collecting all future books and articles on this subject for the publication of a bibliography covering the next decade, i. e., 1970–1980.

2.2. Special investigations

2.2.1. The economics of exchanges

Following a proposal made by Unesco *I. Gombocz* conducted a thorough investigation on the basis of replies to a questionnaire from 32 exchange centres and 240 institutions in 53 countries. The main conclusion of his report on the "Economic aspects of the international exchange of publications"[15] were the following:

- the majority of exchange centres perform transmission functions;
- exchanges promote the exchange of science and culture;
- it is indispensable for materials outside the book trade;
- exchange is more economic in the case of currency difficulties;
- practical steps would make exchanges more effective and less expensive: standardization of correspondence and of balancing methods, reduction of postal and customs charges, etc.

Two special questions raised in this basic paper were made subjects of further investigations. In two studies on the forwarding function of national exchange centres *P. Genzel*[16] compares the actual costs of transmission through an exchange centre with the (fictitious) direct dispatch of the same number of postal units, and comes to the conclusion that central transmission in its entirety is less economic than direct dispatch. Depending on various factors prevailing in different countries, central transmission in collective consignments is therefore not always the most economic method of dispatch of exchange materials, and particularly in the case of periodicals, not always satisfactory.

Taking up a former examination again, *M. Razumovsky* made "A further inquiry into methods of establishing the value of publications received by and sent on international exchange."[17] She describes existing practices in equating this value, and outlines the difficulties with the so-called "priced exchange". All methods aim at overcoming currency problems, reducing work, and satisfying both partners. The author recommends the application of different methods to different exchange materials, to use the priced exchange cautiously because of changing book prices and to make more use of the page-for-page method. Since the problem has kept its topicality, M. Razumovsky has undertaken a new study on "Rising book prices and the exchange of publications".[18] The problem was to learn the reaction of libraries to price developments – in fact one of the crucial problems of today's acquisition and exchange policies. The result was most interesting and encouraging in that the majority of libraries do not reduce their exchanges, but hold an optimistic view as to the future of exchanges.

As far as the exchange of duplicates is concerned, it is often the only possibility of acquiring older literature which is out of print. But it has also an economic aspect. At the 1977 Brussels meeting a case study prepared at the British Library Lending Division was distributed and briefly presented.[19] It is restricted to periodicals and one of its results is that it is economic to handle reasonably long runs. "A costing analysis of the Gift and Exchange Section of the British Library Lending Division" by *E. Tatterton, K. King* and *A. Allardyce* is based on this internal report.[20] This study contains a thorough analysis of each individual working process involved in the handling of incoming and outgoing duplicates and their offer on lists. On the same occasion *H. Rovelstad* presented a detailed survey of the operation of USBE entitled "The economics of the Universal Serials and Book Exchange (USBE)".[21] Particular attention was paid to cost studies. An investigation similar to the one carried out at the BLLD was recently conducted by *R. Ejlersen* at the Institut danois des échanges internationaux de publications scientifiques et littéraires (IDE): "The economic aspect of the exchange of duplicates. Time studies on books. A case study".[22] It is a clear presentation of a complicated problem, and proof is furnished that by introducing certain measures a higher efficiency of work is achieved without lose of substance.

With regard to these three studies it is of course difficult to collect conclusive figures, since different working methods are applied in different exchange centres. Nevertheless, the papers offer a basis for similar investigations which might result in an improvement of duplicate processing.

2.2.2. Developing countries

The Section has already paid great attention to the exchange of publications with the developing countries. Six papers read at the above-mentioned European conference were devoted to this important problem, among others: *R. A. Christophers* on the English-speaking African countries, *J. Lethève* on the French-speaking African countries, *F. Vanwijngaerden* on Central Africa, and *N. R. Einhorn* on South America.[23] The reports on Africa showed the possibility and the difficulties of concluding exchange agreements, but they also made it clear that exchange with African countries is usually more efficient than purchase. The Library of Congress tries to obtain official publications through bilateral agreements with the South American countries, and non-official publications through direct exchange relations with individual organizations, libraries and institutes thereby encountering also some difficulties.

F. Vanwijngaerden in an article on "Improving exchanges of publications with developing countries in Africa. A few suggestions"[24] takes the problem up again and in his latest paper on the "Exchange of publications with developing countries" he draws some general conclusions.[25] One of them is that owing to different conditions no practical recommendations can be given which are applicable to all developing countries in the same way. The author analyses the different problems facing librarians in these countries and pleads for a code of ethics and mutual generosity on which to base all exchanges with the third world.

In this connection it must not be left unmentioned that *P. Genzel* and two of his colleagues prepared a paper on "The realization of the UAP principle" for the IFLA Seminar for Librarians from Developing Countries (Potsdam, GDR, 22–26 August 1978).[26] It deals with the acquisition and exchange of domestic and foreign literature and national and international interlending. Although special emphasis is laid on the problems of the German Democratic Republic, there are frequent references to the developing countries and in the chapter on exchange a few suggestions for its improvement have been made.

2.2.5. Exchanges and interlending

There is no doubt that interrelations as well as differences exist between the two ways of supplying literature which, at a given time, is not available in a library. At the beginning of the European conference on the exchange of publications this problem was briefly dealt with, by among others *D. J. Urquhart.*[30] From the papers and the discussions thereafter, it was clear that loans and exchanges supplement each other, but as yet rarely in any systematically organized way.

The theme was resumed soon after by *A. Sterescu,* in whose library international loans are operated by the exchange centre, in her paper on "The interrelation between the international exchange of publications and interlibrary loan".[31] Most of the answers to the questionnaire which formed the basis of the investigation showed that exchange centres consider loan as a completely separate activity, in no way related to exchanges. In some centres, however, certain aspects of coordination have already been put into practice. Institutions or libraries which are not primarily preoccupied with exchanges combine the two activities.

The problem was given further attention in a basic paper by *A. Allardyce* and *S. C. J. Vickers* on "The international exchange of publications and international lending – is there a useful relationship?"[32] This interesting problem was already on the agenda of a joint meeting of the Sections on Exchanges and Interlending at Štrbské Pleso 1978, and after all a well functioning loan and a smoothly running exchange service are the foundations of UAP. The authors considered the question from two points of view: the substitution of one activity for the other and activities of a supplementary nature. One of the conclusions is "that the principal relationship at present between international lending and exchange is in the form of either a gift for the benefit of the borrowing library or an offer to exchange for the benefit of the supplying library". Although it is not feasible to lay down exact principles, at least some rules of thumb can be derived as to substitution, supplementing and preference given to either exchange or interlending according to the special situation.

3. Final remarks

From our report it follows that on the one hand a good many projects have already been realized, but that on the other hand much is still to be done.

As already mentioned two projects are in train, i. e., a study on the effects of international exchange conventions and the bibliography on the international exchange of publications for the period 1970–1980 (which will be included in the next 5-year programme).

Other new projects have also been suggested.[33] For example: the problem of the inclusion of reprographic materials in exchanges is to be extended to the application of modern technology in general, i. e., the impact of automation on exchange practices. With a view to a comparative study the Section will also be concerned with national legal regulations for the exchange of publications[34] as a special branch of library legislation at large. Another investigation concerns the problems of in-service training of exchange staff in libraries (professional requirements, training programmes).

Some of the projects can and must be carried out by the Section on the Exchange of Publications on its own resources. But there are also others which require joint action with other Sections and Divisions of IFLA. As regards the Division of Collections and Services, new signs of a common approach to handling important problems reaching beyond the individual Sections are already recognizable. In this way an even better contribution can be made to UAP.

References

* Revised and enlarged version of a Working Paper presented at IFLA General Council Štrbské Pleso (1978).

** Director, Acquisitions Department, Deutsche Staatsbibliothek, Berlin, German Democratic Republic

1 Medium-term Programme of the Committee on the Exchange of Publications. In: IFLA Journal, vol. 1, no. 4, 1975, p. 318–320. See Genzel, Peter. On the Medium-term Programme 1975–1980 of the Committee on the Exchange of Publications. Working Paper, IFLA, General Council, Lausanne, 1976. 10 p.

2 Gombocz, István. Forty years of the Committee on the Exchange of Publications. Rev. by Maria J. Schiltman. In: IFLA Journal, vol. 1, no. 1, 1975, p. 9–19.

3 Schiltman, Maria J. (ed.). The international exchange of publications: proceedings of the European conference held in Vienna from 24–29 April 1972. Pullach/München, Verlag Dokumentation, 1973. 135 p. See also Schiltman, Maria J. Report on the European Conference on the International Exchange of Publications held in Vienna April 24–29, 1972. Working Paper, IFLA, General Council, Budapest, 1972. 7 p.; Gombocz, István. European conference on the international exchange of publications, Vienna, 1972. In: Unesco bulletin for libraries, vol. XXVII, no. 1, 1973, p. 54–56; Genzel, Peter. Europäische Konferenz über den internationalen Schriftentausch. Ein Tagungsbericht. In: Zentralblatt für Bibliothekswesen, vol. 86, no. 9, 1972, p. 537–541.

4 In: Schiltman, Maria J., l. c., p. 20–27.

5 In: Unesco bulletin for libraries, vol. XXVI, no. 3, 1972, p. 141–149.

6 First presented as Working Paper, IFLA, General Council, Brussels, 1977. 23 p.
7 Ed. by Frans Vanwijngaerden. 165 p. As no. 4 of the series Documentation, libraries and archives: bibliographies and reference works. French version: Manuel des échanges internationaux de publications. Spanish version: Manual del canje internacional de publicaciones.
8 Allardyce, Alex. Letters for the international exchange of publications. A guide to their composition in English, French, German, Russian and Spanish. Ed. by Peter Genzel. Pullach/München, Verlag Dokumentation, 1978. 148 p. (IFLA Publications 13.) See also Genzel, Peter. Letters for international exchange. Working Paper, IFLA, General Council, Brussels, 1977. 6 p.
9 In: Unesco bulletin for libraries, vol. XXV, no. 5, 1971, p. 282–284.
10 La standardisation des formulaires dans la pratique des échanges de publications. In: Schiltman, Maria J., l. c., p. 114–116.
11 In: Schiltman, Maria J., l. c., p. 37–46.
12 First presented as Working Paper, IFLA, General Council, Brussels, 1977. 23 p.
13 In: Schiltman, Maria J., l. c., p. 49–62.
14 1974. II + 63 p., published by the Belgian Commission on Bibliography in its series Bibliographia Belgica as no. 127. See also the review by P. Genzel in IFLA Journal, vol. 2, no. 1, 1976, p. 58–59, moreover, Kanevsky, B. P. La bibliographie sur les échanges internationaux de publications. In: Schiltman, Maria J., l. c., p. 117–119.
15 In: Unesco bulletin for libraries, vol. XXV, no. 5, 1971, p. 267–281.
16 The efficiency of collective consignments. In: Schiltman, Maria, J., l. c., p. 66–73; The efficiency of the transmission function of national exchange centres for the international exchange of publications. In: Unesco bulletin for libraries, vol. XXX, no. 2, 1976, p. 83–89 and 106.
17 In: Schiltman, Maria J., l. c., p. 74–77.
18 First presented as Working Paper, IFLA, General Council, Copenhagen, 1979. 5 p.
19 Steemson, R. J. A study of costs at the British Library Lending Division's Gifts and Exchange Section. 1977. 4 p. + 3 appendixes.
20 First presented as Working Paper, IFLA, General Council, Štrbské Pleso, 1978. 14 p.
21 First presented as Working Paper, IFLA, General Council, Štrbské Pleso, 1978. 5 p.
22 First presented as Working Paper, IFLA, General Council, Copenhagen. 1979, 13 p.
23 In: Schiltman, Maria J., l. c., p. 80–103.
24 In: Unesco bulletin for libraries, vol. XXX, no. 2, 1976, p. 90–92.
25 First presented as Working Paper, IFLA, General Council, Copenhagen, 1979. 17 p.
26 Genzel, P., G. Meyer, E. Rother. Die Verwirklichung des Prinzips der Universalen Verfügbarkeit der Bestände (UAP). Working Paper, 1978. 22 p. Also in English.
27 M. Zehrer in: Schiltman, Maria J., l. c., p. 104–105.
28 See footnote 5.
29 First presented as Working Paper, IFLA, General Council, Brussels, 1977. 6 p. + appendixes IA–XIB.
30 Exchanges and inter-library lending. In: Schiltman, Maria J., l. c., p. 30–33.
31 Working Paper, IFLA, General Council, Grenoble, 1973. 18 p.
32 Working Paper, IFLA, General Council, Copenhagen, 1979. 15 p.
33 Some new initiatives come from two recent papers by S. C. members of the Section on Exchanges in the jubilee number of IFLA Journal (IFLA and contemporary library problems; vol. 3, no. 3, 1977): Allardyce, A. Some opportunities for IFLA's Division of Collections and Services (p. 243–246); Kanevsky, B. P. IFLA and certain problems of library theory (p. 257–262).

34 An example for a regulation entirely concerned with exchanges is the decree promulgated in the GDR in 1976. See Genzel, P. A decree for the exchange of publications issued in the German Democratic Republic. In: IFLA Journal, vol. 3, no. 1, 1977, p. 84–85.

International Book Exchange of National Libraries in the Late 70s

B. P. Kanevsky*

The exchange of publications between libraries of different countries represents one of the most available and widely spread forms of cultural cooperation and this is why the necessity of expanding it is mentioned in the Final Act of the Helsinki Conference of 1975 a number of times. IFLA and its Section on the Exchange of Publications devote their efforts to the development of book exchange on a regular basis.

National libraries (NL) hold an important place in the general book exchange system.

In 1977 the State Lenin Library (USSR), in agreement with the IFLA Section on the Exchange of Publications, collected data on the state and prospects of international book exchange (IBE) in national libraries (NL) of many countries. These data put us into the picture on the state of affairs in this field and this information has not lost its significance to this day. Many features which characterize the state of book exchange of national libraries in the late 70s will continue to be valid in the early 80s too.

A questionnaire was sent to the NL and national centres of IBE in 80 countries. It contained 26 questions dealing with international book exchange and related problems drawn up in such a way that they did not so much stimulate the gathering of statistical data as reveal the opinion of specialists from different countries on the main trends in the development of book exchange. Answers were received from 38 national libraries and 5 institutes and IBE services from 19 European countries (Austria, Bulgaria, Czechoslovakia, Denmark, FRG, Finland, France, GDR, Great Britain, Hungary, Iceland, Italy, Norway, Portugal, Sweden, Switzerland, USSR, Yugoslavia and West Berlin), 6 African countries (Algeria, Arab Republic of Egypt, Democratic Republic of Madagascar, Rwanda, Senegal, Tanzania) and two countries of North America (USA and Canada) and one country of South America (Brazil).

The first questions of the questionnaire dealt with the production of books and journals, as publications of this type make up the material basis of IBE. Answers to this questionnaire confirmed the tendency towards reduction of the rate of growth of the number of titles of books and journals published annually, a tendency which made itself known by 1977. In a number of countries, after two decades of incessant rise in these indices, stability and, in some cases, even reduction of these publications were observed. Unesco statistics are also indicative of the fact that in 1975 these data proved to be lower than in the preceding year, a factor noted for the first time in many years. For example in the FGR the output of books (in titles) dropped by 15,5 %[1]. Nevertheless, within the framework of the general tendency towards the deceleration of the rate of growth of the number of published titles in general, the production of books which are most frequently used in IBE continued to increase in the majority of countries. These books include original scientific and technical works, official publications and materials of international and national conferences. One can, on the basis of this information, draw the conclusion that the material, book basis of IBE is not diminishing, but potentially growing. From the answers to the questionnaire it follows that the main sources of acquisition of books for IBE are: legal deposit (45 %)[2], additional purchasing of books and subscription to journals (43 %) and the purchasing of books, surplus of the book trade at reduced prices (12 %).

The majority of national libraries (28 out of 41 or 68 %) have at their disposal special collections of publications intended for IBE purposes. National libraries in 14 countries do not have such collections. The number of books kept in these exchange stocks fluctuated from 9 titles of periodicals and 35 titles of other publications in the NL of Malaysia and one hundred copies of books in the University Library in Rwanda to many tens of thousands of volumes in the NL in the USA, USSR, GDR, Bulgaria, etc. According to rough estimates stocks used for IBE run to 1 mill. 300 thousand copies of books, serial publications and journals. Precise calculation, it should be mentioned here, proved impossible because of the difference in the methods of stocktaking, but on the basis of information received from 27 NL we determined that on the average national libraries send out annually no fewer than 1,153,000 copies of printed matter, including 400 thousand books through IBE channels. If we take into account the fact that these statistics also cover literature (including periodicals) which is sent abroad directly by book trade firms and organizations, then we can assume that the bulk of publications held in book exchange stocks is not used. (According to our estimates at least 12 % of publications remain unused.) Apart from this, the comparison of information received from the NL with UNESCO statistics showed that a considerable part of the book and journal output of each country does not find its way into the stocks for exchange of national libraries and thus remains outside the IBE framework. Experience of many years standing in the field of IBE shows that to this category of printed matter belong mainly expensive books and journals produced by commercial publishing houses.

The character of publications exploited in IBE is revealed in the answers from 40 libraries and book exchange centres: 14 of them (35 %) use not only publications intended for sale, but official and free publications of scientific societies, universities and other institutions and organizations, too. 9 NL (22,5 %) exchange only official publications, 8 (20 %) use only publications which appear on the book market. 6 libraries (15 %) utilize both official publications and printed matter intended for sale and, finally, 3 libraries (7,5 %) deal only with free publications. On the whole, the prevailing tendency is to use literature of different kinds in the sphere of book exchange, though the growth of prices on the book market naturally impedes the wider use of commercial printed matter by libraries.

Books are used especially actively in book exchange by the NL. Of the 35 NL, which answered the appropriate question of the questionnaire, 24 (68,6 %) considered their exchange to be active, 9 (25,7 %) described it as being moderate and only 2 libraries (5,7 %) said it was poor.

In the exchange of serial publications the appropriate indices are 54,5 %, 33,3 % and 12,2 % while in the exchange of journals the figures are 72,2 %, 16,7 % and 11,1 %. Newspapers are used to a much lesser degree (20,8 %, 41,7 % and 37,5 % respectively). 27 NL sent us information on the exchange of duplicates out of which 11 libraries (40,7 %) rated it as good, 9 (33,3 %) as moderate and 7 (26 %) as poor.

We marked the slack development of exchange of microfiches, microfilms, photocopies, printed music, maps, posters, musical records and tapes. Only 3 NL said that their exchange of microfilms was active, three other libraries rated it as moderate. Not a single library uses microfiches or microcards either actively or moderately; only 6 NL use photocopies moderately. Only one library exchanges printed music, maps and posters actively, while three other libraries carry on a moderate exchange of these materials. The majority of libraries did not answer these questions at all and only a few libraries indicated that their exchange of such materials was extremely passive.

IBE is usually conducted on the basis of a non-monetary equivalent and out of the 42 libraries that exchange new books, 21 (50 %) conduct a volume-for-volume exchange, 8 (19 %) a page-for-page exchange and 13 (31 %) use the money equivalent. In the exchange of new periodicals the ratio of the non-monetary equivalent (title-for-title) and monetary equivalents is 66 % and 34 %. In the exchange of old books it is 61 % and 39 %; and in the exchange of old periodicals it is 76 % as against 24 %.

Publications received through the book exchange system make up a considerable part of all foreign acquisitions of NL. The answer from 40 NL and book exchange centres supplied the information that these libraries annually receive approximately 865 thousand foreign books through different channels of acquisition. (On the whole, it can be assumed that NL in all countries of the world annually acquire from 1 mill. to 1.5 mill. volumes of foreign books.) Out of this amount over 400 thousand volumes or 46 % were received through the international book exchange system. Differences in the systems of counting periodicals made it impossible to determine the place book exchange holds in the acquisition by the NL of foreign periodicals, but we can assume that here, too, IBE plays an important role.

Not a single library or book exchange centre of the 40 that filled in the questionnaire used computers for IBE purposes, but 6 libraries answered that they were planning to use them in the future. Nevertheless, even if these libraries have realized their intentions, they still make up an insignificant minority.

37 national libraries mentioned difficulties, which impede them in the development of IBE, most of them pointing out more than one obstacle. The most widely spread difficulties were those connected with the acquisition of domestic publications (17 NL), financial difficulties (14 libraries) and shortage ofqualified personnel (11). In other cases the following difficulties were named: lack of space for the housing of literature (6), little interest in foreign books displayed by users (5), difficulties encountered by partners in acquiring and sending the required items (3) and the language barrier (1). These difficulties were mentioned by NL of even such large and highly developed countries as the USA, Japan, France, FRG, Great Britain, GDR, Italy, etc. The only library that said it experienced no difficulties in this field was the National Library in Algiers.

But regardless of these obstacles many national libraries rate high the state of their book exchange: of the 42 NL and book exchange centres that answered the appropriate question 22 (52,4 %) considered their book exchange to be satisfactory, 16 (38,1 %) appraise it as good and only 4 (9,5 %) described it as unsatisfactory.

An optimistic view on the development of international book exchange in the near future (till 1985) was expressed by many experts and taking the opportunity offered by the questionnaire they spoke highly of it, and said that they intended to develop it (NL and book exchange centres of Algeria, Canada, Czechoslovakia, Finland, France, GDR, Great Britain, India, Rwanda, Senegal, Singapore, Sweden, Switzerland, Tanzania, Yugoslavia).

The positive attitude of national libraries to IBE is explained by the fact that this system represents a very important source of acquisition of foreign publications. It is of particular importance to national libraries. The main typological feature of the NL (receiving, keeping and use of the most comprehensive collection of domestic printed matter in the country) determines the universal character of its stocks, including the foreign part. The acquisition of foreign literature in subject fields covered by the NL is either of a universal nature (Library of Congress in the USA, the State Lenin Library, the British Library) or of multi-aspect character (the national libraries of Canada, France, Japan, Belgium and some other countries). The national libraries of the majority of socialist countries and their stocks are also of universal or multi-aspect character (the national libraries of Bulgaria, GDR, Poland, USSR, Czechoslovakia, Vietnam, Cuba and some other countries).

The breadth of acquisition of foreign literature and, as a rule, the comprehensive stocks for exchange in the NL, made up, of among other things, surplus legal deposit copies and non-planned acquisitions, determine their interest in IBE. Comprehensive acquisition of domestic printed matter makes the national library the last and highest instance to which foreign libraries and scientists turn in their search for the publications of the given country which they need.

As a result of this development the NL are not only active partners in IBE (in a number of countries they are the most active participants), in a number of cases they also function as internal book exchange centres or are closely connected with such centres if they happen to be independent administrative units. On the whole, the IBE of a national library plays an important role in the development of the entire national system of book exchange.

Summing up these observations we can draw the conclusion that today IBE of the bulk of national libraries is characterized by the stability and poor use of new carriers of information, microforms in particular. Unused stocks for exchange, consisting of old and new publications, represent certain resources for the development of IBE. Nevertheless, the main obstacles in the way of the progress of IBE are identified with difficulties of a financial order.

The future of international book exchange of national libraries mainly depends on the size of the budgets of these libraries and in these conditions one can expect that the tendency towards a very strict economic policy in the field of IBE will continue to make itself known, but international book exchange will, undoubtedly, remain an important component of the activity of national libraries.

References

* Chief, Department of Foreign Acquisition und International Book Exchange, The Lenin State Library of the USSR, Moscow.

1 Unesco. Statistical Yearbook. 1976. Paris, 1977, p. 802–916.

2 The percentages are estimated on the basis of the general number of answers received by us.

The Multilateral Unesco Exchange Conventions (1958) and Their International Significance

Vladimir Popov*

In a rapidly changing world which has devoted vast resources to penetrate the enigma of the microcosm and the secrets of outer space but which still tolerates some 800 million illiterates, the need for knowledge is one of the basic needs of mankind. And, despite the proliferation of new technologies for the mass media, the printed word remains the principal agent for the dissemination of knowledge. Hence the increasing need for books, magazines, scientific journals, etc. in all parts of the globe and the steady growth of publishing activities throughout the world.[1] The creation in recent years of a multitude of new independent states striving to lay the foundations of their cultural and scientific development has greatly increased the need for printed matter.

At the same time, however, the functioning of the book trade and the acquisition of publications through normal commercial channels is often disturbed and even interrupted by a variety of factors such as economic and political barriers, different levels of economic growth in individual countries, inconvertibility of currencies, differences in investment policies, etc. To this should be added the unhealthy rise in publication prices in recent years, which has become a world-wide phenomenon and has led to further deterioration in international book trade as well as library acquisitions.

In view of all this, libraries in many countries have been compelled to activate and reinforce the international exchange of publications so as to be able to meet the need to extend their collections. The normal way of acquiring library materials is still the book trade of course, but the international exchange of publications has gained a parallel importance for many libraries as a secondary yet important source for library acquisitions. The international exchange of publications is nothing new; libraries in Europe and America have been using it for a long time. Since World War II it has been adopted by libraries in other continents, too.

The international exchange of publications is a product of international confidence and cooperation and is, by its very nature, itself an important factor in stimulating and extending cultural and scientific cooperation among nations. The importance of these exchanges has been firmly established and widely demonstrated during the last 90 years[2], in the following ways:

1. Until recently the international exchange of publications was the only way, and still remains a most important way for many libraries, to obtain foreign official publications and government documents. The growing diversification both in type and content of these materials increases still more the significance of international exchanges in this field.

2. Many important publications of universities, research institutes, learned societies, and the like are available only on an exchange basis.
3. Acquisition through exchanges is the most suitable way for museums and art galleries to obtain the publications of other museums and kindred institutions, especially since they are usually rather expensive.
4. International exchange is for many libraries in a number of countries the main procurement source for materials which would normally be obtained from the book trade. These libraries lack sufficient funds for their acquisitions or belong to countries with inconvertible currency and/or insufficient reserves of foreign currency. But other libraries, too, have recently turned to exchanges, in view of the inflationary spiral that has affected book prices all over the world.
5. Last but not least, international exchanges play an important part in giving publicity to the publications of research institutes, learned societies, and the like. It is through this kind of mutual exchanges conducted by such institutions that their publications can easily be directed to the interested readership[3].

International Regulation of the Exchange of Publications

Considering the importance of the international exchange of publications, its international regulation appears essential. It is commonly accepted of course that informal agreements between exchange partners (libraries, institutes, universities, academies and other institutions) are the best basis for their mutual exchanges. A number of factors operating both at the national and the international level can however have an adverse effect on the international exchange of publications[4], and the influence of these factors cannot be eliminated by informal agreements only. Hence the necessity for regulation on an international level.

The common form of such regulation is represented by the multitude of intergovernmental bilateral agreements specially concluded for exchange purposes or, more often, by the cultural conventions containing also provisions concerning the exchange of publications. Though of considerable importance, these agreements and conventions are not enough. To establish a broad, universal, internationally accepted legal basis which will overcome differences and difficulties arising from the legislation of individual States (customs and administrative restrictions, etc.) and achieve a certain degree of uniformity in such matters as standardization of exchange techniques, methods concerning the settlement of exchange accounts, the nature of exchange objects, the rights and obligations of exchange partners and so on, multilateral conventions are necessary, and this necessity was recognized as early as the end of the 19th century.

There are at present several multilateral conventions on the international exchange of publications. Of these only four are of a universal character: the two Brussels conventions of 1886, and the two Unesco conventions concluded in Paris in 1958 at the 10th General Conference of Unesco.

The Multilateral Brussels Conventions[5]

These are formally still in force, although the Paris conventions ought to have superseded them. The chief merits of the Brussels Conventions may be described as follows:

a) They have been instrumental in giving publicity to the international exchange of publications and in fostering the practice of exchanges.

b) They have established a new and most useful exchange institution, viz. the National Exchange Service (or Centre) as a body to be directly responsible for the exchange operations conducted in each of the contracting States.

c) They have stipulated for the making and circulation of lists of materials available on exchange.

d) They have regulated the methods of transportation of exchange objects (special arrangements to be made in the case of overseas exchange deliveries).

e) Generally speaking, the Conventions blazed a trail for wider cultural cooperation among nations.

After being applied for 70 years, the Brussels Conventions became an object of criticism and almost everybody agreed that they no longer served the regulation and development of the international exchange of publications, and that for the following reasons[6]:

a) The conventions were rigid. They imposed on every party the obligation to operate exchanges with all other contracting parties. Each contracting State was obliged to supply the total of its production of official publications to all other partners. This was a heavy obligation, and a most unfair one to countries with a large production of official publications, as they could not obtain from countries with a small production of such publications even an approximate equivalent of what they had supplied. This prevented the conclusion of many bilateral agreements.

b) Official publications were defined rather vaguely as objects of the exchange. As a result, a great many contractors formed their own opinion on this subject and insisted on it.

c) National exchange services (or centres) were granted only limited powers; they were mere mediators and executive bodies and could not influence the conclusion of bilateral agreements nor decide on the objects to be earmarked for exchange.

d) The Conventions did not take into account possible future developments, they did not foresee and could not cope with the enormous extension both of publishing activities throughout the world and of international cultural and scientific cooperation; nor did they provide for the development of exchanges for technical documentation and scientific and bibliographic materials.

e) They did not impose on the contracting parties the obligations to encourage and facilitate the exchange of publications, to expedite exchange deliveries, and to grant postal and customs concessions to exchange services.

f) There was no clause in the conventions providing that they must be systematically amended, brought up to date, and attuned to the ever-changing conditions on international dealings.

For these reasons, the conventions were ratified by comparatively few countries during a period of 90 years following their conclusion. Convention A was ratified by 21 countries; Convention B by 17. A number of big countries, such as France, USSR, Germany, Great Britain and Japan, did not accede to them. These conventions, then, never, became a universal regulator of the international exchange of publications.

The Paris Unesco Conventions concerning the International Exchange of Publications

After a preparatory period lasting several years[7], the 10th session of the Unesco General Conference held in Paris in 1958 adopted two multilateral conventions on the regulation of the international exchange of publications, namely: Convention concerning the International Exchange of Publications, and Convention concerning the Exchange of Official Publications and Government Documents between States. By their structure and by the flexibility of the formulations adopted, the conventions are a decisive step towards the consolidation of exchange practices and are quite unprecedented.

The formulations of this convention are of a rather general character and adumbrate a wide scope for the international exchange of publications, providing at the same time the general legal basis for its development and further regulation at an international level. (The Convention concerning the Exchange of Official Publications and Government Documents between States is on the other hand considered a special arrangement.)

1. Stimulation and facilitation of the exchanges. Article 1 contains the most important text disclosing the object of the Convention. It imposes a common, explicitly formulated obligation on the contracting States: 'to encourage and facilitate the exchange of publications between both governmental bodies and non-governmental institutions of an educational, scientific and technical, or cultural nature, which are non-profit making in character, in accordance with the provisions of the present Convention'. This obligation refers both to the exchange of materials published or subsidized by the governments themselves and to all other forms of exchanges. It rests with the governments of the contracting States to decide on the forms of facilitation and stimulation of the exchange; it suffices that they do not contradict the provisions of the Convention. Only some of these forms are expressly indicated in the Convention: the establishment or employment of national exchange services or centres; paying of packaging and transmission costs; granting favourable rates and transport conditions as well as exemption from customs duties, etc.

   It is evident that the governments of the contracting States have been assigned a major role by the Convention and that their attitude towards international exchanges is of the utmost importance to the development or the retardation of the exchanges[8].

2. Exchange objects. Article 2 enumerates the objects that may be considered appropriate articles to be exchanged, on condition that they shall not be resold, i.e. shall not be used for commercial purposes. These are publications of an educational, legal, scientific and technical, cultural and informational nature, such as books, newspapers and periodicals, maps and plans, photographs, microcopies, musical works, Braille publications and other graphic material; also official publications and government documents (as specified by the special Convention). This broad formula includes book trade publications, too, which may also be exchanged (subject to the above-mentioned condition that they shall not be resold). Confidential documents and similar materials which, for any reason, have not been made public, are expressly excluded from the exchange. Gramophone records and magnetic tapes are also excluded[9].

3. Institutions operating exchanges. In the sense of art. 4 international exchanges may be operated by governmental bodies (acting as persons of the civil law only, and not as exchange authorities in the sense of the special convention[10]), learned societies, laboratories, cultural institutions – on condition that they are non-profit organizations.

4. National exchange services and centres. As organizations with the specific task of supporting and facilitating the international exchange of publications they were created for the first time by the Brussels Conventions. National exchange services exist at present in some 50 countries. The establishment of the majority of them was not based on the multilateral exchange conventions. There are different opinions about the scope and legal nature of the functions of the national exchange services and even some doubts about their usefulness, yet the raison d'être of these institutions has been fully justified by their activities.

   Art. 3 gives a flexible interpretation of the duties and functions of the national exchange services and centres. The main objective of these organizations is the development and coordination of the exchange of publications[11]. To fulfill this task the exchange centres perform various functions, such as transmit exchange material abroad or from abroad[12]; supply advice and information on exchange matters and, more specifically, on exchange possibilities for domestic and foreign institutions; stimulate (when it is considered appropriate) the exchange of duplicates; collect statistical information on the country's international exchanges and submit statistical reports to the pertinent governmental bodies thus enabling them to prepare reports to Unesco on the working of the Convention; conclude direct informal agreements with other centres or participate in the negotiation and conclusion of governmental bilateral exchange agreements, etc.

5. Bilateral exchange agreements. According to art. 12, the contracting States may conclude, whenever it is deemed necessary, bilateral exchange agreements 'for the purpose of supplementing the present Convention and regulating matters of common concern arising out of its application'. It should be noted, however, that the Convention retains its full legal validity without the conclusion of bilateral agreements.

6. Support and international coordination by Unesco. To assist Unesco in implementing effective international coordination of the exchange of publications, the governments of the contracting States are bound to send Unesco annual reports on the application of the Convention and copies of the bilateral agreements entered into in accordance with art. 12. On the basis of these reports and additionally collected information Unesco is obliged to prepare and publish studies on the application of the Convention and also to publish information on the activities of the contracting States in this field.

Upon request made by the governments of the contracting States, or on its own initiative, Unesco may give financial and technical assistance in connection with the establishment and organization of national exchange centres and also in connection with any problems that might arise from the application of the Convention.

Convention concerning the Exchange of Official Publications and Government Documents between States

This convention is a model of a sound and sensible international agreement avoiding the faults of a number of previously concluded similar conventions. It is flexible and sufficiently mandatory at the same time. It shows a great resemblance, both in structure and content, to the general Convention, but differs widely in some features.

1. Exchange partners. Exchange partners are the States themselves, not libraries or other institutions (art. 1). Actually, the exchange of official publications is conducted by the national exchange services (centres), or, where no such services exist, by a central authority specially designated for the purpose (art. 4). No government may entrust the functions of exchange to any other institution. It follows that the governments retain immediate control over this type of exchanges. Therefore, the national centres have to agree with the relevant governmental bodies a modus operandi in implementing these exchanges and obtain sanction for the conclusion and contents of bilateral agreements with foreign centres. It is the national exchange service centre or the central authorities specially designated for the purpose that are responsible in each country for the application of the Convention. They are given, accordingly, the powers required to obtain the material to be exchanged and sufficient financial means to perform the functions of exchange (art. 4, § 2).
2. Bilateral agreements. According to art. 3, the contracting States enter into bilateral agreements for the purpose of implementing the Convention and regulating matters of common concern arising out of its implication. However, there is no obligation to conclude bilateral agreements (except in cases where the country's legislation expressly requires this). It is quite possible for the partners to operate exchanges of official publications solely within the framework of the Convention. They can thus avoid the protracted and sometimes ponderous procedures of concluding bilateral agreements[13].
3. Exchange objects. Art. 2 enumerates official publications and government documents but this listing is not exhaustive. The contracting States are

entitled to decide on the categories of official publications to be exchanged and define, on the basis of mutual agreement, the content of the term 'reciprocity' contained in art. 1. On the basis of such agreements the national exchange centres will make lists of the materials to be exchanged and fix their number. These lists and the number of official publications designated for exchange may be modified by arrangements between the exchange centres concerned.

The remaining texts of the Convention are identical with those of the general Convention.

Ratification of the Conventions

The Conventions were opened for ratification or acceptance (according to the constitutional procedures of individual States) by Unesco Member States and for accession by non-member States. They came into force in 1961 after the third ratification document for each of them was registered with the Director General of Unesco. According to recent information kindly supplied by Unesco, each of the Conventions has been ratified or accepted by 38 States[14].

Despite numerous appeals for accession to the Conventions made at various international conferences, seminars and other meetings convened by Unesco and IFLA on problems of the international exchange of publications, the number of States adhering to the Conventions remains comparatively small considering Unesco's aspiration to create a universal system of international exchange of publications.

It is worth mentioning that adherence by countries of different continents is far from being balanced. The Conventions have received very strong support in Europe; 21 European States have already ratified both Conventions. And it is unquestionable that nearly all European countries with well-developed library systems and active international exchanges have adhered to the Conventions. It is also interesting to note that some of the European countries that still abstain (such as Switzerland, Yugoslavia, Portugal) have acceded to the Brussels Conventions.

Unlike Europe, little support has been given by other continents. There have been 7 ratifications from North and South America, 4 from Asia, 5 from the Pacific area. It is most discouraging that only very few of the developing countries have adhered to the Conventions so far. The core of the developing countries is still absent.

The main reasons for abstention seem to be the following:

- Authorities and librarians in a number of countries, especially in the new States, may still be unfamiliar with the text of the Conventions as a result of insufficient publicity by Unesco, IFLA, the national library associations and the library press[15].
- The still rudimentary development of the economy, of scientific research and of cultural activities in some countries; this may account for the insignificant demand for foreign literature and documentation, and hence for the little interest in the Conventions as factors facilitating foreign acquisitions.
- Misjudgment of the advantages that adhering countries can gain from the Conventions; lack of understanding or underestimation of the great significance of a universal system of international exchanges for the development of education, science and culture in all countries.
- Inherited mistrust in multilateral international regulation of the international exchange of publications as a result of the faults and failure of the Brussels Conventions.
- Absence of interest in the Unesco Paris Conventions because of the previous accession of some countries to regional conventions on the international exchange of publications (e. g. the interamerican conventions) the stipulations of which possibly satisfy, even to a considerable extent, the needs of those countries. Therefore, they may consider it quite useless to join and ratify the Unesco Paris Conventions.

It should be borne in mind, however, that mere ratification is not an end in itself. The Conventions do not automatically produce an effect upon ratification. Everything depends on the good will and understanding, on the far-sightedness of the authorities in a country, on their firm determination to promote the country's exchanges, on the will to build up a resourceful national system for the international exchange of publications as an integral part of an effective universal system of exchanges[16]. It is obvious, then, that the implementation of resolutions II.1, II.3, and II.5 of the European Conference on the International Exchange of Publications held in Vienna in 1972[17] is of particular importance as this would strengthen national exchange centres and provide better conditions for exchanges in all countries. The number of ratifications of both Conventions almost doubled in a decade or so. It can now be safely claimed that the Unesco Paris Conventions are the first international exchange agreements to have gained true universality.

The International Significance of the Conventions Today

The conventions are of no less importance today than they were twenty years ago, since the general conditions that have generated them continue to exist:

- Many countries have only recently begun to develop their exchange of publications.
- In a substantial number of countries legal regulation has not yet reached sufficient proportions.
- Publishing activities all over the world keep growing at an ever-increasing speed; publisher prices grow much faster than currency allocations to libraries.
- Owing to the sensational progress of science, technology and economic development, the need for foreign scientific materials also becomes ever more urgent. This is particularly true of materials obtainable on exchange only.
- To stimulate development of the economy, science and technology, education and culture, the new States in Africa and Asia have to maintain strong and reliable scientific and cultural contacts with the old world.

As Donald F. Wisdom put it some 10 years ago, "The Agreement on the Importation of Educational, Scientific and Cultural Materials signed in Florence in 1950 and the Convention concerning the Exchange of Publications (1958), being the most important stages in the history of international co-operation performed under the auspices of Unesco, both facilitated large-scale exchange of publications among States and opened up new perspectives for the fulfilment of the exchange programme"[18]. This is particularly true today. The Paris Conventions have proved a most useful instrument in the hands of Unesco in promoting international cooperation in the field of education, science and culture.

The Conventions help to bridge the gap between less developed and advanced countries with respect to the distribution of educational, cultural and scientific resources. They assist Unesco and many governments in combating illiteracy and in implementing a variety of programmes, e. g. for youth and adult education, for the advancement of science, for free dissemination of knowledge and information, etc. They have been instrumental in carrying out the library and information programmes of Unesco. Hundreds of exchange agreements have been reached along the lines of the Conventions, and millions of books, periodicals, documents, etc. have been exchanged under the terms of those agreements.

The Conventions proclaim a number of internationally acceptable general principles which provide a sound basis for the regulation of the exchange of publications. Even libraries in non-ratifying countries follow these principles in establishing exchange contacts with partners abroad. The Paris Conventions have thus developed into some sort of common law in the matters of exchange all over the world and have become a charter of international interlibrary cooperation.

With a view to attaining some of the principle objectives of Unesco, a number of major interrelated international projects have in recent years been developed within IFLA, FID, ICSU/AB, and Unesco, namely: UNISIST, UBC and NATIS. They have already been made operational as parts of the Unesco General Information Programme in which IFLA is closely cooperating. A new component of particular importance to the library world has been added recently to this integrated system: the ambitious IFLA project of the Universal Availability of Publications, supported by Unesco, which has been gaining momentum and has already been incorporated into IFLA's medium-term programme[19]. The core of this project is the establishment of a universal interlending system of cooperating national lending systems (including national libraries and other major libraries) possibly aided by several supranational libraries in possession of comprehensive collections. They will all have one principal goal: to provide free international access to the information resources of all nations by making available all existing publications through loan or photocopy. The fulfilment of this ambitious but extremely difficult programme presupposes, of course, a long period of very active acquisition of such foreign materials as may be required to build up the appropriate national and supranational collections, which would be unthinkable without new and extended large-scale exchange programmes for all the libraries involved in the implementation of a world-wide system such as UAP. This would inevitably require new efforts on the part of the governments of participating countries as they would have to assess the efficiency of national library systems in respect of the international exchange of publications and take the appropriate measures to stimulate and extend exchange activities. This, in turn, would again focus attention on possible accession to the multilateral Unesco Conventions and on making the best possible use of the opportunities provided by the Conventions to contracting States. In short, the Conventions must be considered to be a valuable and indispensable instrument in promoting the UAP system, and this will undoubtedly add incentive to adhering to them.

How Can a Country Benefit from Adherence to the Convention?

It is not possible to enumerate here all the advantages of adherence to the Conventions. Nevertheless, it is worth trying to describe briefly the most important of them:

1. All libraries operating international exchanges are given a new status. By the act of ratification or acceptance, each country proclaims its willingness to

give full support to libraries in organizing a system of exchanges (e. g. by establishing one or several exchange centres, by providing them with sufficient funds and official publications for exchange purposes, by granting them exemption from customs duties and favourable treatment on postal and transport rates, etc.). As the prestige of existing exchange centres grows they get more support, and their work gains in efficiency; they can more easily overcome bureaucratic procedures and other impediments[20].

2. The ratification of any multilateral convention usually leads to the conclusion of bilateral agreements between States repeating the main principles and stipulations of the multilateral convention. However, as international practice shows, libraries in countries which have already ratified the Unesco Conventions very often arrive at direct informal agreements by mere reference to the Conventions. This often applies even to the exchange of official publications.
3. Ratification provides excellent opportunities to libraries and other institutions to extend their international exchanges of publications or initiate exchange projects. Both Conventions are of particular importance to libraries which are not entitled to approach foreign institutions without referring to similar international arrangements. The exchange of official publications may be a case in point.
4. Ratification provides exchange centres and libraries with the legal ground for exerting pressure on the administrative bodies controlling them and even on high authorities, with a view to obtaining better regulation of exchange activities, more financial support, more favourable treatment with regard to customs duties, transport rates, etc.
5. The activities of national exchange centres and services performed in accordance with the Conventions (e. g. collection and publication of statistical data and information on exchanges, coordination, etc.) enable the relevant governments to form a better and more precise idea of the country's exchanges and to regulate them better.
6. Ratification opens a prospect of obtaining from Unesco under the provisions of art. 10 or 12 of the Conventions substantial financial and technical assistance in connection with exchange activities, which may involve help in establishing national exchange centres and organizing a dependable system of exchanges, training of exchange personnel abroad, obtaining consultations from foreign experts, and so forth. This is of particular importance to new States and developing countries as they could be enabled to build up national exchange centres of their own with the generous aid of Unesco.
7. The Conventions should particularly appeal to developing countries. They could benefit in many different ways from accession to the Conventions, e. g.:

a) The establishment of international exchange contacts gives new-born libraries in these countries the status of members of the world community of libraries and other cultural and research institutions. At the same time it enables them to obtain from all the exchange partners abroad many valuable publications, badly needed by their countries in connexion with their economic, scientific and cultural development.

b) The flow of exchange materials from libraries in developing countries to exchange partners all over the world enables other countries to gain an insight into their problems, needs and achievements and to become familiar with the specific features of the individual development of each. As a result, universities and research institutes in other countries may not only be more readily prepared to share experience with developing countries, but also to conduct exchanges on more generous terms and often to donate valuable publications to their libraries, universities and other cultural institutions.

c) International exchange of publications will ensure wide circulation and publicity for the works of distinguished authors, scholars, researchers and scientists working in developing countries and help them to establish most useful personal contacts with colleagues abroad.

d) Exchange personnel in developing countries will be able to use training facilities in foreign countries more extensively.

e) Libraries in developing countries will be entitled under the provisions of the Conventions to call upon the relevant governmental bodies for full support in matters of exchange and to seek priorities in building up library collections.

f) The Conventions provide for better and more extensive acquisition of materials by libraries, information centres and archives thus contributing to the establishment and development of a resourceful national information system along the lines of Unesco's NATIS concept.

Conclusion

The most important and indeed the crucial factor in the campaign to raise the number of adhering States is the attitude of national library associations, national libraries, and, in general, the attitude of the whole library community in each country. The importance of concerted action aimed at persuading a country's government to take the necessary steps towards adherence cannot be emphasized enough. This action would require to be backed up with evidence to show the benefits that could be derived from adherence.

Librarians should use their energy and experience to help Unesco with the implementation of its General Information Programme and to support its efforts at promoting the dissemination of knowledge and cultural values throughout the world. Campaigning for adherence to the conventions is one of the best ways of doing this.

References

* Deputy Director, National Library Cyril and Methodius. Sofia, Bulgaria

1 Delavenay, Emile. Pour le livre. Paris, Unesco, 1974, p. 8.

2 Cox, William. The new exchange conventions. Unesco bulletin for libraries, vol. XV, no. 4, 1961, p. 171–177.

3 Ejlersen, Rita. Decentralization of exchange services. In: Conférence sur les échanges internationaux de publications en Europe. Budapest, 13.–19.9.1960. Budapest, Centre bibliothéconomique et méthodique de la Bibliothèque Nationale Széchényi, 1962, p. 112.

4 Hamel, G. A. The international exchange of publications. Some barriers. In: Conférence sur les échanges internationaux de publications en Europe, op. cit., p. 189–200. See also: Kanevsky. B. P. The international exchange of publications and the free flow of books. Unesco bulletin for libraries, vol. XXVI, no. 3, 1972, p. 141–149.

5 Convention A concernant les échanges internationaux pour les documents officiels et pour les échanges scientifiques et littéraires, conclue à Bruxelles le 15 mars 1886.

Convention B pour assurer l'échange immédiat du journal officiel ainsi que des annales et des documents parlementaires, conclue à Bruxelles le 15 mars 1886.

For details see: Handbook on the international exchange of publications, 4th ed. Paris, Unesco, 1977. Chapter IV, p. 66; Chvatov, V. J. Voprosy regulirovanija knigoobmena mezhdu bibliotekami i nauchnymi uchrezhdeniiami raznih stran. Leningrad, Biblioteka AN SSSR, 1963, p. 33–35; Gombocz, I. Inquiry into the execution and results of Unesco's exchange conventions 14 years after their adoption. In: The international exchange of publications. Proceedings of the European Conference held in Vienna from 24 to 29 April 1972. Ed. by Maria Schiltman. München, Verlag Dokumentation, 1973, p. 37.

6 Petersen, E. N. Steps towards the adoption of new conventions for the exchange of official and non-official publications. Libri, 1958, No. 4, p. 296–308.

Chvatov, V., op. cit., p. 42.

Warsaegger, C. Multilateral conventions concerning the international exchange of publications. Unesco bulletin for libraries, vol. XVII, no. 2, 1963, p. 53–62.

Gombocz, I., op. cit., p. 38.

7 Dargent, J. L. Les nouvelles conventions internationales d'échanges. Bruxelles, Commission belge de bibliographie, 1960. 270 p.; Handbook, op. cit., p. 66; Chvatov, V., op. cit., p. 42–46; Wærsegger, C., op. cit., p. 58–59.

8 Kanevsky, B. P. The international exchange of publications. Unesco bulletin for libraries, vol. XIX, no. 6, 1965, p. 303.

9 Wærsegger, C., op. cit., p. 60.

10 Cox, W., op. cit., p. 175.

11 Details on the organization, functions and the legal nature of the national exchange services (centres) see in: Avicenne, P. The mission of the national exchange services. Unesco bulletin for libraries, vol. XVIII, no. 6, 1964, p. 253–258; Handbook, op. cit., p. 99–113; Slabczyński, W. New trends in the international exchange of publications. Unesco bulletin for libraries, vol. XXII, no. 5, 1968, p. 218–224 and 235; Popov, V. Short surveys of the European exchange centres. In: The international exchange of publications, op. cit., p. 49–50.

About the coordination of the exchange of publications, its nature and organization, see details in: Kovács, M. Coordination des activités nationales d'échange. In: Conférence sur les échanges internationaux de publications en Europe, op. cit., p. 128–141; Jabloński, Z. The role of the international exchange of publications in the library system. In: The international exchange of publications, op. cit., p. 24–25; Chvatov, V., op. cit., p. 58–68.

12 It is not possible now to maintain that the transmission function of exchange centres is a function of primary importance, especially after the economic analyses of exchange carried out by István Gombocz and Peter Genzel in the last decade. Genzel has shown that centralized transmission is by no means more economical than direct delivery and cannot be always recommended as the most efficient method of forwarding exchange material. See Genzel, P. The efficiency of collective consignments. In: The international exchange of publications, op. cit., p. 66–73; Genzel, P. Die ökonomische Rentabilität der Weiterleitungsfunktion einer internationalen Austauschstelle. Zentralblatt für Bibliothekswesen, vol. 86, no. 6, 1972, p. 321–331; Genzel, P. The efficiency of the transmission function of national exchange centres for the international exchange of publications. Unesco bulletin for libraries, vol. XXX, no. 2, 1976, p. 83–89 and 106. See also: Slabczyński, W., op. cit., p. 222.

13 Cox, W., op. cit., p. 173.

14 The Convention concerning the International Exchange of Publications has been ratified or accepted by the following countries (in alphabetical order): Belgium, Byelorussian SSR, Brazil, Bulgaria, Czechoslovakia, Cuba, Denmark, Dominican Republic, Egypt, Ecuador, Finland, France, German Democratic Republic, Federal Republic of Germany, Ghana, Guatemala, Hungary, Indonesia, Iraq, Israel, Italy, Luxemburg, Libya, Malawi, Malta, Marocco, The Netherlands, New Zealand, Nigeria, Norway, Panama, Poland, Rumania, Spain, Ukrainian SSR, USSR, United Kingdom, USA.

The Convention concerning the Exchange of Official Publications and Government Documents between States has been ratified or accepted by: Belgium, Byelorussian SSR, Bulgaria, Central African Republic, Czechoslovakia, Cuba, Denmark, Dominican Republic, Egypt, Ecuador, Finland, France, German Democratic Republic,

Federal Republic of Germany, Ghana, Guatemala, Hungary, Indonesia, Iraq, Israel, Italy, Luxemburg, Libya, Malta, Marocco, The Netherlands, New Zealand, Nigeria, Norway, Panama, Poland, Rumania, Spain, Sri Lanka, Ukrainian SSR, USSR, United Kingdom, USA.

15 The Conventions have so far been published in: the documents of the 10th session of the Unesco General Conference in 1958; two separate brochures in 1958; the UNO Treaty Series for 1958; the Unesco bulletin for libraries, vol. VIII, no. 2–3, 1954, p. 2 ff; the 3rd and the 4th editions of the Handbook on the international exchange of publications. They are also published in full in Dargent, J. Les nouvelles conventions internationales d'échanges, op. cit. Their text is also included as an annex in the book of V. Chvatov (see ref. 5 above). However, all this does not seem to be sufficient.

16 See: The international exchange of publications, op. cit., p. 43.

17 Ibid., p. 133.

18 Wisdom, D. American Libraries and the Unesco programme of library development. Unesco bulletin for libraries, vol. XX, no. 5, 1966, p. 268.

19 IFLA. Medium-Term Programme. The Hague, 1976. p. 23–26. Line, Maurice. Universal availability of publications. Unesco bulletin for libraries, vol. XXXI, no. 3, 1977, p. 142–151.

20 Genzel, P. Europäische Konferenz über den internationalen Schriftentausch. Zentralblatt für Bibliothekswesen, vol. 86, no. 9, 1972, p. 538.

Exchange of Publications with Developing Countries

Frans Vanwijngaerden*

Introduction

I have been asked to write a paper about the exchange of publications with developing countries. My credentials for talking about this topic consist of more than fifteen years of work in the field of exchanges and especially of exchanges with developing countries since the European Conference on the international exchange of publications held in Vienna in 1972, where I delivered a paper about my experience in exchanging publications with the Republic of Zaïre.

There already exist of course a number of books, articles in journals and papers submitted to professional conferences and seminars on the exchange of publications in general and on the exchange of publications with the developing countries in particular. The scope of this paper is not to open new horizons, but to offer some suggestions.

I have had no opportunity to conduct a survey or to send out questionnaires asking the opinions of colleagues in different countries about the possibilities, the practicalities and the situation of the exchange of publications. My paper, therefore, is based in the first place on my personal experience of many years and in the second place on the readings of articles, studies and research reports and papers presented to various professional seminars and conferences.

I have tried to avoid the pitfall of becoming lost in reminiscences and in detailed descriptions of personal experience. My aim has been to analyse problems systematically and to draw meaningful conclusions. However, I must confess I did not find the endeavour easy and I am not sure of complete success.

Importance of Exchange

In order to establish the importance of exchange as a means of acquiring library materials I will give some figures from three of the world's great libraries and quote three authoritative librarians.

In 1975 the Library of Congress acquired 489,893 pieces through exchange. In 1977 the British Library got 440,345 pieces through exchange and the figures of the Lenin State Library are respectively 60 % of foreign books, 88 % of serials and 82 % of periodicals. In my own library approximatively 40 % of the total acquisitions are made by exchange.

A glance at these figures shows clearly that however limited it ought to be in theory, in practice exchange takes a very big place.

G. E. Morrison of the British Library's Department of Oriental Manuscripts and Printed Books[1] states that "acquisition by exchange is particularly important for certain categories, such as official government publications, and academic serials; and also in relation to certain countries where it proves to be the most convenient, or even the only possible method of acquiring library materials, since it overcomes the difficulties of currency control and exchange; and in some cases because all publication, even of cultural material, is done by government agencies".

Enid Bishop of the Australian National University[2] tells us that, "although most Australian libraries seem to favour selective purchasing, exchange is a very fruitful potential source for acquiring publications; a source which has not yet been fully exploited. She found that about 75 % of the Korean publications in the collections of Eastern materials in the Australian libraries were acquired by gift and exchange. For Japanese materials it turned out that exchanges are most useful for a wide range of institutional and government publications and for getting publications not for sale or not readily available through normal trade or official channels".

To end these statements, a quotation from A. Allardyce, head of Gift and Exchange and of Monograph Acquisitions of the British Library Lending Division, Boston Spa, U.K.[3]: "Most librarians would say that, however agreeable it is to have enough money to buy, no major and few specialised research libraries can acquire by purchase all the publications they need. So, despite the acceptance by exchange librarians that as a general rule exchange is no substitute for purchase, they remain convinced that U.A.P. will never be possible without exchange and that progress towards it will be quicker with exchange."

These quotations clearly demonstrate the importance of exchange in acquiring library materials.

No rose without its thorn

A. Allardyce suggests[3] that pronouncements like those of Alexandre Vattemare, an enthusiastic pioneer of international exchange, and of Unesco, in defining the aim of international exchange make it sound like a world mission: "to place at the disposal of every nation all the sources of intellectual culture... which are produced by the most favoured of its fellows" (Vattemare, 1864); "Maintain, increase and diffuse knowledge by... encouraging cooperation among the nations in all branches of intellectual activity including... the exchange of publications... by initiating methods of international cooperation calculated to give people of all countries access to the printed and published materials produced by any of them" (Unesco's Constitution 1948). His own view is that: "... like other roads to UAP, exchange of publications is not a broad highway, but a multitude of pathways and byways which connect individual libraries. It is not a self-contained operation but a part of one phase, the preliminary acquisition phase, of making the world's literature available to readers."

As well as having limitations exchanges have unsatisfactory aspects, too, such as e.g.:

1. the failure of some exchange partners to send serial issues regularly:
2. the difficulty in establishing satisfactory reciprocity in the exchange relationship, where there is difficulty in estimating the value of publications sent and received,
3. the limitation of libraries to meet the requests of exchange partners for trade books, owing, as A. Allardyce points out[3] to the "growing needs of their own libraries in relation to book funds shrinking in real terms and unfilled staff vacancies, caused by restrictions on public expenditure".

Exchange with developing countries

Despite the thorns in the acquisition through exchange, I am, however, convinced that exchange between industrialized and developing countries is now more necessary than ever.

In developing countries, the demand for scientific and other kinds of publications is becoming more and more urgent, because of their desire to build up a modern economy and develop their national self-awareness.

Owing to the complexities of the social and economic conditions of developing countries, and their level of development in library services, which vary from one country to the other, it is difficult to say definitely what is practical and what is possible in the planning and the establishment of exchange programmes for the whole developing countries.

A further point made by A. Allardyce[3] is that: "A developing country with shifting priorities, as well as a great power in social and industrial revolution, may leave its libraries with little alternative to exchange for much of their foreign acquisitions. This is because national considerations favour other import priorities."

Developing countries: A Correction

This paper is specially concerned with the problem of exchanges of publications with developing countries. It is interesting to speculate on the possibilities of improving these exchanges. But before doing so, I should like to correct a misconception. There is a tendency to regard all developing countries as poor countries, but this is a mistake. Many are certainly impoverished, but some still at a relatively early age in their library development are rich.

On the other hand we should not forget that several of the countries classified as "least developed" by the United Nations have established library services comparable in many respects with those which exist in much wealthier countries.

Let me also bring to your attention a basic problem outlined by Dudley Sears and Leonard Joy in their survey of world development problems[4]: "We refer to some countries of the world as 'underdeveloped' or 'developing', but we must realize that these descriptions are dangerous evasions of reality. All countries are 'underdeveloped' in various senses, and the most serious problem about the countries usually described as 'developing' is precisely that few of them are successfully doing so."

Perhaps the most important point to be made at the beginning of this paper is that, as the Committee for Development Planning of the United Nations Economic and Social Council (ECOSOC) has pointed out[5] "while developing countries as a group face more or less the same general problems of underdevelopment, the difference between the poorest and the relatively more advanced among them is quite substantial" and "the least developed among them cannot always be excepted to benefit fully or automatically from such general measures adopted in favour of all developing countries".

Exchange: A code of Ethics

"There no longer exist privileged centres where science is cultivated in isolation. In all countries, among all races, there are achievements of the human brain that are contributing to scientific progress.

Intelligentsia are seeking each other, they group themselves, human solidarity is steadily widening and the spirit of partnership, after having changed the world of economics, is working now new wonders in the world of the mind.

Every day in academies, scientific institutes, learned societies, special committees, literary societies, editorial committees, joint effort is creating new sources and inexhaustible manifestations of the human brain. In our times, one of the quickest ways of hastening the development of science and the onward march of progress is surely to collect these products of the human brain, centralize all the various efforts, draw up in each country an inventory of the general progress of intellectual work, place at the disposal of inquiring minds what might be called the dossier on each subject, prepared and supplemented by the specialists and the scientists from the new and the old worlds, bring them into association and, so to speak, into collaboration with each other...".

I cannot restrain from quoting this passage from the report of the Belgian International Exchange Commission to the Belgian House of Representatives on the occasion of the ratification on 19 January 1887 of the Brussels Conventions of 15 March 1886. Despite the passage of time, these words have lost none of their original significance.

They clearly state the basic rule of all exchange of publications, which must above all be based on a code of ethics, on mutual generosity.

Problems facing librarians in developing countries

All of us are aware of the difficulties which the developing countries face in the procurement of library materials. They are well summarized by J. A. Ombu when he wrote[6]: "Amongst the many problems that face librarians in the developing countries, problems relating to the acquisition of foreign materials seems to be the most acute. These problems exist mainly for three reasons: first the relatively underdeveloped economic circumstances of the developing countries within which their libraries have to function; second, the fact that book publishing is in its infancy throughout most of the area, and therefore the necessity to buy almost entirely from foreign countries; and the fact that the developing countries are far removed from the metropolitan areas of the world, where the bulk of the publishing is done."

There are, however, more problems which developing countries have to face; problems which are obstacles often standing in the way of exchange, and even of the distribution of publications:

1. There is a *serious lack of trained personnel* in the right proportions at various levels in order to provide a satisfactory infrastructure. The problem

of staff is for instance raised in the report of the Expert Meeting on National Planning of Documentation and Library Services in Africa, organized by Unesco at Kampala (Uganda) from 7 to 15 December 1970. The conclusions include recommendations to the effect that, as regards staff training, a sufficient number of fellowships should be granted to student librarians who would be sent for training to library schools. Furthermore it was suggested that the courses run at universities should be reshaped so as to lead to a qualification equivalent to either a first or a second degree in librarianship, according to what type of library personnel best meets the country's stated requirements.

In order to meet these requirements exchange librarians in developing countries must be imaginative, should be familiar with the basic sociological, psychological and economic realities of their countries, must determine the nature and quantity of library materials that are needed, and need constantly to improve their competence to identify, select, acquire and utilize effectively the best library materials for their readers. This of course is an idealistic view. Even allowing for the greater tempo of life today, and the rapid strides made in the developing countries, we of the so-called developed countries should not expect our younger colleagues to run before they can walk. On the contrary, we should be less concerned with the full stops and commas of bureaucracy and more with awareness that the growth of a living and vital democracy in any country must (in the conditions of the modern scientific-technological world) depend on its access to the ideas of the whole world.

2. A second general problem, characteristic of developing countries, is *a basic lack of general communication facilities* (telephone, mail services, etc). One letter I received on this subject reports that "periodicals and publications rarely reach their destination, they are either sent to another address or disappear on the way". A. J. Loveday summarized it[7]: "In the development of any coordinated national scheme for acquisitions it will be important to give due weight to the considerable difficulty which exists in most developing countries over the whole range of communications. First, postal services tend to be organized in relation to box numbers associated with institutions rather than physical locations of individual addresses and this can present problems for the individual researcher or reader in remote rural places depending upon a distant central back-up service... The postal system in the majority of cases is slow and does not support the rapid availability of photocopies between acquisition centres. Telecommunications too may be sporadic and unreliable bases for cooperative procedures. It may frequently be easier and more rapid to obtain inter-loan on an intercontinental rather than on an inter-regional or national basis."

The removal of such obstacles requires action, often on a large and expensive scale, by government authorities. Librarians of the developed

2.2.3. Official publications

Questions of official publications and government documents played an important part in the former Committee on the Exchange of Publications, the former Sub-Committee on the Exchange of Official Publications and, for a while, the Committee on Official Publications established in 1963/64. Joint meetings of the Committees on Exchange and on Official Publications also took place, e. g., in 1972. This tradition will be continued, whenever problems of common interest are to be solved.

At the European conference on the exchange of publications *"The union catalogue of African official publications"* comprising 1186 entries from 38 African countries at that time, was commented on.[27] Some trends of the exchange of official publications can be seen through an analysis of the questionnaires assembled in 1970 by the same author and published in the collection "Questionnaire-Sammlung der IFLA-Kommission für Amtsdruckschriften". The results confirm that, in the majority of cases, national libraries and centres have sufficient scope for the further expansion of their exchanges of official publications. In his above paper *B. P. Kanevsky*[28] also gives a detailed analysis of this project and stresses the urgency of including official publications in exchange, since they do not only contain legal, economic and political material but also often scientific and technological information as well.

2.2.4. The role of reprography

In inter-library lending reprography (photocopies, microfilms, etc. instead of originals) has already played a major role for many years. In his paper on "The inclusion of the products of reprography in the international exchange of publications" *N. R. Einhorn*[29] presents the results of a questionnaire received from 338 institutions throughout the world on the present use made of these materials in their exchange programmes (instead of originals, e. g., if a work is out of print, or in case of single copies of journals). The result was rather disappointing and can be considered a proof of the conservative attitude of many exchange librarians towards new techniques. Only about one-sixth of the institutions do use reprography in exchange and less than 10 % regard reprographic materials as an important part of their exchange programme. On the other hand, the use of these materials in exchange has been steadily increasing since 1950, and the rate of increase has more than doubled since 1960. Microfilm and photocopy are predominant among the materials used. More research is needed on the technical specifications used in the production of these materials and in the exchange bases used by the respondents.

countries can do no more in most cases than draw attention to their adverse effects on library development. We surely know that cooperation in librarianship, as in other areas, is one of the most difficult activities man can undertake, and expectations should not be high. In an imperfect world there will be a gap between hope and realization. But we should not forget that it is primarily an ethical problem concerned with a professional's duty to help those less well provided for than himself. It was perhaps best expressed by T. J. Minder[8]: "The first factor of co-operation is that of individualism, or the ego. The individual member of a co-operative rightly expects some personal gain from the venture...

The second factor is attitudes or the willingness to relinquish one's personal gain for the benefits of the other participants... The third factor is group benefits."

3. A third general problem characteristic of developing countries is that there is frequently *insufficient external currency* to provide the basic needs of a modern information and library service or there is such bureaucratic delay involved in the procedures of obtaining external currency that the materials required come too late or do not come at all (e. g. because they are sold out or out of print).

   Here the industrialized countries can play a vital role. The monetary crisis has had repercussions on the economies of almost all countries in the world and even the governments of the industrialized countries have been led to enter upon a period of austerity in their policies. But after all libraries in the developed countries are relatively larger and have richer resources and are consequently able to help their colleagues in the developing countries to obtain valuable foreign publications.

   Our colleagues in the developing countries should, however, have in mind that, although it has certainly been accepted that the industrialized nations ought to lend their cooperation to the developing countries and promote their advancement in every way possible, it remains obvious that the remedies for development lie mainly in their own hands and those of their own authorities.

   They should not forget that the progress of a nation depends chiefly on its own will to work, its spirit of initiative, and its effective determination to reach the targets it has set itself.

   Finally efficient, continuous co-operation from outside cannot be found, indeed cannot be requested of the international community, that is to say principally the industrialized countries, unless the governments and the responsible authorities of those countries are able to convince their public opinion and their parliaments of the useful nature of this policy, and to do so by exhibiting the efforts already being made by the countries which are to receive it and which are ready and willing to make the best possible use of it.

In short one must not try to compare the internal effort with the outside contribution just if they were contradictory factors, but rather consider them as complementary to one another, the latter being the indispensable extension and reinforcement of the former.

In relying on their own strength our colleagues in the developing countries will learn that in this spirit exchange programmes will in the end be valuable to all parties involved.

Librarians in developed countries should also remember that all publications sent on exchange must be carefully selected. There is indeed a very natural tendency to increase quantities. For technical reasons, young libraries are not yet in a position to assimilate an unduly large number of additional publications; if they attempt to do so, quality is in danger of being sacrified to quantity. Accordingly, one should keep permanently in touch with the exchange partner to ensure that both parties avoid sending publications which are not desired.

And here again librarians in the developing countries must be aware of the important role they play. They ought to remember always that a good library service has tremendous potential for the economic and social good of the people and that in the developing world where resources – human, material and financial – are scarcest, its perfection is a necessity.

4. A fourth general problem, characteristic of developing countries rather than industrialized countries is that there is *an inadequate indigenous book trade infrastructure* to provide the expertise required for an efficient book purchase service for libraries, information centres, etc.

   The reasons for the poorly developed document production and distribution system in some developing countries have been pointed out by J. S. Parker[9]: "The local document production system – broadly speaking, authorship, printing and publishing – may be poorly developed, partly because low educational standards produce few authors, partly because a limited potential user population does not provide a large enough market, at least to support commercial publishing... Even when the production side is well-developed – which is rarely the case in developing countries – the local document distribution – essentially bookselling, but including also legal deposit provision and other means of distribution – may be ineffective, making it difficult for libraries to obtain the documents they need, even when they know them to have been produced."

The industrialized countries should have these facts in mind when establishing their exchange programmes with developing countries, especially now that world financial crises tend to affect exchange programmes adversely. I have been informed that more and more libraries are reviewing their exchange agreements so as to reach a near parity in the materials exchanged. We should, however, be aware that we are serving a noble code of ethics and accordingly base our exchanges on a certain amount of generosity and should not mind when often the pendulum of giving and receiving swings in the direction of the receiving library. A librarian should not be a conservator, a curator, he should be the exploiter of information par excellence.

Even if only a few publications are received in return, it should be remembered that these may acquire lasting historical value.

It should also be remembered that it has long been the practice of national libraries and similar bodies to arrange for the exchange of all or selected publications of their home government, with similar publications from other countries, and this has proved to be a very effective form of collection building. In view of the wide range of subjects covered by official publications, similar arrangements between institutions other than national libraries may be a fruitful way of increasing literature coverage, particularly for those countries where problems of foreign exchange exist.

Conventions concerning the international exchange of publications adopted by Unesco in 1958

The librarians of the developing countries should stimulate their governments to accept and take into consideration in cultural agreements the two Unesco Conventions of 1958 on the International Exchange of Publications and on the Exchange of Official Publications and Government Documents between States, adopted by the General Conference at its tenth session, Paris, 3 December 1958.

These two Conventions are the results of a long evolution. In the beginning exchange experts paid little attention to bi- or multilateral conventions. The opinion prevailed that direct contacts were much more effective and that the smooth flow of international exchange of publications was best served by direct and informal agreements between libraries, institutions, universities, academies, etc. However, the idea of utilizing multilateral conventions gained ground, and in 1977 the Section on the Exchange of Publications submitted during the 43rd IFLA Council Meeting at Brussels the following resolutions:

Recognizing the importance of the international exchange of publications as an aid to achieving universal availability of publications (UAP) IFLA invites Unesco:

1. to remind the governments of the member states of the reciprocal facilities available under the Exchange Conventions;
2. to ask them to consider signing Conventions where they have not already done so;

3. at the same time and in support of 1, to circulate to these governments literature illustrating the advantages of adhesion e.g. a pamphlet containing the texts of the Conventions and also a description of their relevance, based on the paper by V. Popov presented at the World Congress of Librarians in Brussels[10];
4. to publish a version of this paper in the Unesco Bulletin for Libraries and elsewhere if appropriate;
5. to conduct a study, in conjunction with the Section on the Exchange of Publications and possibly with the proposed IFLA Office for UAP, on the practical effects of the Exchange Conventions since 1958 on progress towards UAP.

The texts of the two Unesco Conventions can be found in the Handbook on the International Exchange of Publications.[11].

The Need of a National Publications Exchange Centre

It should be pointed out that it is useful to set up a national publications exchange centre, at least where such a centre does not exist already. This would help us to attain the fundamental objectives laid down in Unesco's Constitution, which enjoins us to "maintain, increase and diffuse knowledge... by initiating methods of international co-operation calculated to give the people of all countries access to the printed and published materials produced by any of them".

I think that the role of an exchange centre should be entrusted to the national library. The British report on university libraries published in 1967 gave a comprehensive list of the possible functions of national libraries, and pointed out that it ought to be i.a. the centre of the country's international exchange of official publications and that a national centre for the distribution of duplicate material was useful in ensuring that surplus material was screened to prevent valuable books being lost to the nation[12].

I know that our colleagues of the national libraries in the developing countries have to deal with many problems, the principal being the creation of a central body to plan and oversee the integration and co-ordination of the activities of various types of library services in order to provide everybody with equal opportunity of access to the total information resource of the country.

It is however very important that the national library should also assume general responsibility for exchange of publications, facilitate such exchanges between libraries, institutes and learned bodies and distribute foreign publications received in consignments, which are not addressed to particular readers.

In doing this it should develop its own exchange policy in association with other libraries of the country, and not seek to draw off all the talent to its own services. Before all the national library's policy must form part of the nation's overall information policy.

We in the industrialized countries may not forget that our colleagues in many African countries, and developing countries in general are aware of the urgent need there is to consider seriously ways in which their efforts could be harmonised and co-ordinated to avoid wastage, undue duplication and the resultant inaffectiveness in services.

They have besides to deal with many other problems, such as:

1. the promotion of education, the fight against illiteracy, etc. They are very frequently required, therefore, to undertake the tasks of a public library and of a national library at the same time;
2. the necessity of effecting transfer of knowledge and information by using modern technology to record and transmit the original message to the next generation; a necessity arising from the fact that in some developing countries the intellectual records of society are not necessarily in graphic form, but in oral and visual.

We should therefore never forget that in the developing countries, where thousands of homes are without water or sanitation, let alone electricity, telephones or television, we must take a realistic view and base our contacts firmly on an understanding of just how serious these obstacles are and just how difficult they are to overcome. A special effort should consequently be made to give the national library the means to perform the functions of a national exchange centre, which has now become a very complex matter.

Handbook on the International Exchange of Publications

In order to help colleagues in developing countries, Unesco charged the IFLA Section on the Exchange of Publications with the preparation of a revision of the Handbook on the International Exchange of Publications[13].

The first edition (1950), edited by J. L. Dargent (Belgium), was a contribution by Unesco to the rehabilitation and development of libraries after the Second World War, and a confirmation of the importance of the exchange of publications for the free flow of information and ideas among nations. It was the first directory of exchange possibilities on a world-wide scale ever published.

The second edition (1956), edited by Gisela von Busse (Federal Republic of Germany), aimed at improving the exchange and distribution of publications available for exchange. The character of a directory was kept in Part Two, while the new Part One contained a practical guide on handling the international exchange of publications in its various forms.

The third edition (1964) enlarged and revised, also edited by Gisela von Busse, followed the arrangement and layout of the second edition, including information on the two Conventions and guidance on the organization of a national exchange centre. It was published as an attempt to cover the increased production of scientific publications and to meet the evergrowing desire on the part of scientists and of governments for exchange of experience in this field.

The fourth edition (1978) aims at providing:

> a guide on the methodology, organization and management of the international exchange of publications; a detailed up-to-date directory of exchange centres with a national responsibility. Details of publications offered for exchange by individual institutions, which appeared in the three first editions, have been omitted from the fourth edition, since they can never be complete or fully up-to-date. Instead they are specially marked in The World of Learning[14].

The Handbook on the International Exchange, which is published in English, French and Spanish editions, is also intended for the use of librarians who want to set up an exchange centre, especially in developing countries. About the 4th edition see also p. 13.

Letters for the International Exchange of Publications

Since libraries are faced with a rising demand for older out-of-print and for new non-trade publications as well as foreign books for which foreign currency may be scarce or unavailable, the scale of exchanges and of the correspondence involved is increasing. It is also more and more accepted that librarians need to communicate with foreign colleagues in languages they understand.

For this reason the IFLA Section on the Exchange of Publications prepared a letter-writer in five languages under the title of "Letters for the international exchange of publications. A guide to their composition in English, French, German, Russian and Spanish"[15].

This booklet contains the phrases essential to the writing of exchange letters. The Section also hopes that their guide can be regarded as a modest contribution both to standardization in international librarianship and to improvement in the exchange of publications between the more developed and the hitherto less developed countries. For further details see also p. 13.

Improving exchange through international cultural relations

In Belgium many departments are actively involved in what I will call international cultural relations. These relations are no longer restricted to organizing an exhibition, a theatrical performance or a concert abroad. They have evolved from this narrow artistic sphere to embrace varied forms of interaction and co-operation in a number of areas, such as educational problems and teaching, artistic and literary culture, socio-cultural activities, information, scientific co-operation, aid in development. Such interaction and co-operation take the form of cultural agreements, bilateral engagements between States in which Joint Commissions decide on the programmes within the framework of a specific financial undertaking.

In this field we librarians can suggest to the responsible department how to act effectively when it has decided to act, for example in improving exchange of publications with developing countries.

I, of course, know that not all countries are tied by cultural agreements. I also know that the scope of cultural action is not restricted to the content of the agreements. Interaction and co-operation can thus take place unilaterally and freely, on the sole initiative of a single country, unconnected with any equivalent reciprocal arrangements. And here again we can do something to recommend the department in charge, that funds will be made available to meet the cost of exchange of publications with developing countries.

In Belgium, as well as in other countries we have a whole range of areas and a serried network of relations involving all the vital sectors of society, making them of direct concern to all members of the government and to a number of public administrations. These international cultural relations have grown progressively in breadth and content. Similarly, the allocation of work has evolved gradually along pragmatic lines, according to whether this or that Department witnesses the affairs coming under its responsibility gradually being drawn into the sphere of international exchanges.

As you see there are several ways which can be used to improve exchanges with developing countries. But unfortunately a large proportion of Belgium's foreign cultural relations is in the hands of various bodies which are not coordinated in any way, each of which conducts its own policy independently and from its own narrow standpoint. Here again we, exchange librarians, should inform the departments involved what can be done in order to organize suitable contacts with foreign cultures and provide a vital stimulus in the process of cultural cross-fertilization which, basically, constitutes the real significance of all international exchanges. We should not forget that foreign cultural co-operation requires, today more than ever, a dynamic and carefully thought-out policy. And in this domain exchange experts can help in bringing about a coherent policy that enhances the cultural image of their country.

The Belgian Foreign Affairs Department conducts some direct activities abroad, though these are restricted to traditional artistic events organized on a small scale and to a few exceptional ventures designed to emphasize, via our diplomatic and consular posts and our cultural attachés, that Belgium maintains a certain permanent cultural presence in the world.

In this way publications are sometimes put at the disposal of libraries in developing countries. The result, however, is most of the time negative, because the executive set-up of these direct cultural activities has in most circumstances not been sufficiently planned and these activities do not help in pursuing a coordinated policy of cultural activities abroad. The coordinating role can perhaps be played by the National Exchange Centre in each country concerned.

We have in Belgium a National Unesco Commission, which was set up in 1948, but which unfortunately no longer corresponds to present-day needs and is consequently almost unable to fulfil its function, which is to act in an advisory capacity to the Belgian delegation attending Unesco General Conferences.

At present the six Ministries involved in Unesco affairs: the Foreign Minister, the two Ministers for National Education, the two Ministers for Culture and the Secretary of State for Scientific Policy are trying to restructure the National Unesco Commission.

In order to help this restructure in a small way the Belgian Service for International Exchange is trying to convince the Belgian Unesco Commission to defend before the Unesco General Conference the resolutions from the Vienna Conference[16]. In this way it is hoped that the exchange of publications with our partners in the developing countries can be improved.

Conclusion

We have been warned, and rightly so, that what we now call the explosion of world literature is really but a splutter, and that the real explosion is still to come. With this in mind, and with the rapidly increasing costs of books and periodicals, librarians everywhere must relentlessly seek every opportunity to cooperate in the developing of the exchange of library materials.

It would also be advisable that a generous attitude should be adopted towards the international exchange of publications and that whenever possible, special funds should be earmarked for the purchase of publications for developing countries.

The developing countries are aware that as they go more deeply into their own culture, they come to appreciate foreign culture better. They are continuing to make great efforts to improve the situation and are fully aware of the problems and short-comings which are still hindering their cultural action.

Since knowledge has no boundaries, all countries should try to make the knowledge in their territories available to the rest of the world. It must be emphasized again and again that one of the ways of making one's own culture more widely known is to carry out regular and judiciously elected exchanges.

This paper has of course not the ambition to be complete. It only gives a few more suggestions than those already given in my papers presented at the Conference of European Experts on International Exchanges in Vienna (24–29 April 1972), at the 38th IFLA General Council Meeting in Budapest (28 August – 2 September 1972) and at the 44th IFLA General Council Meeting in Štrbské Pleso (27 August – 2 September 1978). These are only suggestions which I can offer from my experience. Other and better suggestions may occur to others.

References

* Head, Service belge des échanges internationaux, Brussels

1 Morrison, G.E. Acquiring eastern materials: the exchange scheme of the British Library. Paper presented at the IFLA Worldwide Seminar, May 31 – June 5, 1976. Seoul, Korea.

2 Bishop, Enid. Acquiring eastern material: Australian solutions. Paper presented at the IFLA Worldwide Seminar, May 31 – June 5, Seoul, Korea.

3 Allardyce, Alex. U.A.P. and the exchange of publications. IFLA Journal, vol. 4, no. 2, 1978, p. 122–128.

4 Sears, Dudley, and Joy Leonard. Development in a divided world. Harmondsworth, Penguin Books, 1971, p. 9.

5 United Nations. Economic and Social Council. Committee for Development Planning. 7th Session, 22 March – 1 April 1971. Report. New York, U.N., 1971. (E/4990), p. 12.

6 Ombu, Jigekuma A. Acquisition problems in developing countries. International Library Review, vol. 9, 1977, p. 83–93.

7 Loveday, Anthony J. Some thoughts on the University Library's contribution to collaborative resource sharing in developing countries with special reference to acquisitions. In Resource Sharing of Libraries in Developing Countries. Proceedings of the IFLA/Unesco Pre-Session Seminar for Librarians from Developing Countries. Antwerp University, August 30 – September 4, 1977. Ed. by H.D.L. Vervliet. München, K.G. Saur, 1979 (IFLA Publications 14) p. 48–49.

8 Minder, T.J. Organizational problems in library co-operation. Library Journal, vol. 95, no. 18, 1970, p. 3448–3450.

9 Parker, J.S. Library resource sharing in developing countries: objectives and obstacles. In Resource Sharing of Libraries in Developing Countries. Proceedings of the IFLA/Unesco Pre-Session Seminar for Librarians from Developing Countries. Antwerp University, August 30 – September 4, 1977. Ed. by H.D.L. Vervliet. München, K.G. Saur, 1979 (IFLA Publications 14) p. 17.

10 Popov, Vladimir. The Multilateral Unesco Exchange Conventions (1958) and their international significance. Paper presented at the 43rd IFLA General Council Meeting in Brussels, 3–10 September, 1977. 23 p.

11 Convention concerning the international exchange of publications. Convention concerning the international exchange of official publications and government documents between states (texts in Handbook on the International Exchange of Publications, 4th ed. Edited by Frans Vanwijngaerden, Unesco, 1978, p. 70–80).

12 University Grants Committee. Report of the Committee on Libraries. HMSO 1961 [The Parry Report], p. 85.

13 Handbook on the International Exchange of Publications, 4th edition. Edited by Frans Vanwijngaerden. In the series, Documentation, libraries and archives: bibliographies and reference works, no. 4. Paris, Unesco, 1978. 165 p. (French edition ISBN 92-3-201466-1; Spanish edition ISBN 92-3-301466-5.)

14 The World of Learning [Annual]. London. Europa Publications.

15 Allardyce, Alex. Letters for the International exchange of publications. A guide to their composition in English, French, German, Russian and Spanish. Ed. by Peter Genzel. Translated by Jacques Lethève, Maria Razumovsky, Boris P. Kanyevsky and Adolfo Rodríguez, Pullach/München, Verlag Dokumentation, 1978. 148 p. (IFLA Publications 13.)

16 The International exchange of publications. Proceedings of the European Conference held in Vienna from 24–29 April 1972. Ed. by Maria J. Schiltman. Pullach/München, Verlag Dokumentation, 1973. 135 p.

The Use of Photoreproductions in International Exchange

Nathan R. Einhorn*

The use of the traditional products of the printing press – books, journals, newspapers – in exchanges between libraries has long been an accepted practice. How widespread, however, is the exchange of photoreproductions – not just microforms but also xeroxes, photostats, and photoprints? This was the question posed by IFLA's Section on the Exchange of Publications at its 1974 meeting in Washington. A questionnaire, designed to elicit information which would be of interest and assistance to all institutions engaging in exchange work, was mailed in 1976 to 628 representative institutions in nearly all countries of the world. These countries are arranged, for purposes of analysis, into six groupings, as follows:

1. African-Asian – all countries in Africa, the Near-East, South Asia, and the Far East.
2. The United States.
3. British – Great Britain, Canada, Australia, New-Zealand, and the West Indies.
4. International Organizations.
5. European – all countries of Europe except Spain and Portugal.
6. Hispanic – Latin America, Spain, and Portugal.

In response to the 628 questionnaires which were sent out, completed questionnaires were received from 338 institutions and organizations, i.e. fifty-four percent of those polled responded to the questionnaire. The questions and the tabulations of the replies follow.

1A. *Does your library use photoreproductions in its exchange program?*

Area	Yes	No	Total	Questionnaires sent	% replied
African-Asian	30	45	75	180	42
United States	17	47	64	93	69
British	18	54	72	104	69
Inter. Orgs.	5	20	25	52	48
European	39	48	87	153	57
Hispanic	4	11	15	46	33
Totals	113	225	338	628	54

1B. *If so, do you consider this practice an important part of your exchange program?*

	Yes	No	% Yes	% No
African-Asian	24	6	80	20
United States	12	5	71	29
British	5	13	28	72
Inter. Orgs.	2	3	40	60
European	21	18	54	46
Hispanic	3	1	75	25
Totals	67	46	59	41

1C. *When was it begun?*

	1950–55	1956–60	1961–65	1966–70	1971–	Unknown or No reply
African-Asian	2	2	0	10	10	6
United States	2	0	3	6	2	4
British	0	4	1	4	4	5
Inter. Orgs.	0	0	0	0	4	1
European	4	9	8	6	8	4
Hispanic	0	0	2	0	2	0
Totals	8 (7 %)	15 (13 %)	14 (12 %)	26 (23 %)	30 (27 %)	20 (18 %)

Of the 338 libraries which replied, only one-third indicated that they used photoreproductions in their exchange programs. Sixty-seven (fifty-nine percent of those which used photoreproductions as exchange materials) considered this practice an important part of their exchange programs and 46 (forty-one percent) did not. These figures do not reflect a widespread interest in exchanging photoreproductions. The interest does appear, though, to be increasing. More than half of the ninety-three who knew when their programs began (fifty-six), dated the start of their microform exchanges after 1966.

The reprographic media seem to be endorsed with the greatest enthusiasm by European libraries. Nearly forty-five percent of the respondents in that area indicated that they used reprographic products in exchange; fifty-four percent of these latter in their turn viewed materials in reprographic formats as important to their exchange program. Not surprisingly, many of the Europeans also indicated that they had been using microforms, xerox and the like in their exchanges for a long time (twenty-one of the thirty-five libraries which stated when they began the practice started before 1966). On the other hand, although only seventeen of the sixty-four respondents from the United States claimed to use reprographic products in exchange, twelve of them (or seventy-one percent) felt this practice was important.

Most interestingly, the African-Asian respondents seconded the Europeans in their regard for reprographic materials (forty percent of the respondents claimed use of them in exchange), while fully eighty percent of these users stated that the reprographic materials were an important part of their program. The preponderance of the African-Asian users of reprography in exchange began this practice after 1966.

2A. *What types of photoreproductions do you use for exchange?*

	Af.-As.	U.S.	Brit.	Inter. orgs.	Eur.	Hisp.	Totals
Film	2	3	3	0	2	0	10
Fiche	1	1	2	2	2	0	8
Xeroxes	5	4	4	1	1	0	15
Photostats	1	0	0	0	0	0	1
Film & Fiche	0	0	0	0	1	0	1
Film & Xeroxes	2	7	5	0	15	3	32
Film & Photoprints	1	0	0	0	0	0	1
Fiche & Xeroxes	1	0	0	0	1	0	2
Fiche & Photostats	0	0	0	0	0	1	1
Xeroxes & Photostats	1	0	0	0	0	0	1
Film, Fiche & Xeroxes	1	1	3	0	0	0	5
Film, Fiche & Photostats	1	0	0	0	0	0	1
Film, Xeroxes & Photoprints	6	0	1	0	5	0	12
Film, Photostats & Xeroxes	2	0	0	1	7	0	10
Film, Photostats & Photoprints	0	0	0	0	1	0	1
Film, Fiche, Xeroxes & Photostats	1	0	0	0	1	0	2
Film, Fiche, Xeroxes & Photoprints	0	1	0	0	0	0	1
Film, Xeroxes, Photostats & Photoprints	3	0	0	0	2	0	5
All types used	2	0	0	1	1	0	4

2B. *Which do you use most often?*

	Af.-As.	U.S.	Brit.	Inter. Orgs.	Eur.	Hisp.	Totals
1. Microfilm	9	5	3	0	19	0	36
2. Microfiche	3	1	1	0	2	0	7
3. Xeroxes	11	11	8	1	14	3	48
4. Photostats	2	0	0	1	0	1	4
5. Photoprints	1	0	0	0	0	0	1
6. No reply	4	0	6	3	4	0	17

The predominance of xerox copies in the replies to the first part of this question, backed up by the preference for it expressed in the second part, is striking. Does it indicate that photoreproductions are not yet fully accepted as substitutes for the printed page? The quick copier's product may be satisfactory for temporary use, but in how many cases is it accepted as the archival equal of information in a printed work that will be added to the institution's permanent collections?

Xerox undoubtedly constitutes the bulk of free-of-charge items supplied on a limited basis by so many of the respondents (see Question 7 below), for the simple reason that it is so quickly prepared and is relatively inexpensive. In such cases, the items supplied (often to serve short-term reference needs) do not require copy of archival grade, and xerox is ideal.

3A. *Do you have separate exchange funds to obtain photoreproductions and other special materials for exchange partners?*

	Yes	No	% Yes	% No
African-Asian	10	20	33	67
United States	3	14	18	82
British	1	17	6	94
Inter. Orgs.	0	5	0	100
European	10	29	26	74
Hispanic	0	4	0	100
Totals	24	89	21	79

3B. *What percentage of your total exchange funds are used to acquire photoreproductions?*

	Af.-As.	U.S.	Brit.	Inter. Orgs.	Eur.	Hisp.	Totals
Very small or insignificant	1	2	3	1	9	0	16
1 %	0	7	6	1	1	0	15
2 %	0	1	0	0	1	0	2
3 %	0	0	0	0	0	0	0
4 %	1	0	0	0	1	0	2
5 %	3	0	1	0	3	0	7
6 %	0	0	0	0	1	0	1
7 %	0	0	0	0	0	0	0
8 %	0	0	0	0	1	0	1
9 %	1	0	0	0	0	0	1
10 %	2	1	0	1	1	0	5
More than 10 %	4	0	1	0	1	0	6
No reply	18	6	7	2	20	4	57

These funding questions were formulated to determine the importance of photoreproductions in the exchange program, in terms of the financial resources devoted to them. Forty-five of the fifty-six libraries which answered the second part of the question spent less than ten percent of their exchange funds on microforms, and thirty-one spent two percent or less.

The low priority of funding bears out the relatively widespread indifference, previously noted, to the use of reprographic products in exchange. Significantly, though, one-third of the African-Asian respondents do budget specially for this purpose (followed by the Europeans), and most strikingly, of the eleven respondents who budget ten percent or more of their exchange funds for reprographics, six were from the African-Asian area. Here again we see evidence of the enthusiasm and significance accorded to reprographic exchanges by libraries in the developing nations. While copying costs are undeniably high, the cost of hard copies, especially of materials that must be bought from foreign sources, is undoubtedly more burdensome for these libraries and the exchange of photoreproductions of whatever format (xeroxes and microfilms are clearly preferred) is therefore highly attractive.

4A. *Does your institution have its own facilities for producing photoreproductions to be used in exchange?*

	Yes	No	% Yes	% No
African-Asian	27	3	90	10
United States	14	3	82	18
British	14	4	78	22
Inter. Orgs.	3	2	60	40
European	37	2	95	5
Hispanic	4	0	100	0
Totals	99	14	88	12

4B. *If so, must you reimburse your laboratory from your exchange funds for the cost of such materials?*

	Yes	No	No Reply	% Yes	% No
African-Asian	11	15	1	42	58
United States	8	6	0	57	43
British	3	10	1	23	77
Inter. Orgs.	1	1	1	50	50
European	13	21	3	38	62
Hispanic	2	2	0	50	50
Totals	38	55	6	41	59

Of the 113 institutions which use microforms in their exchange programs, the overwhelming majority have their own facilities. When one remembers to what extent xeroxes predominated in replies to Questions 2A and 2B above, on which type of photoreproduction was used most often, one may well conclude that the facilities most often owned and used are xerox copiers. It is therefore not surprising that these institutions did not have to reimburse their laboratories for the photoreproductions (xeroxes) they used so frequently.

5. *What is the cost (or price) to your institution for the various types of photoreproductions you use on exchange?*

	Negative film (per frame)	Positive film (per foot)	Fiche	Xerox	Photo-stat	Photo-print
African-Asian	Some 10c a frame, 5c a page; $ 13.00 to $ 40.00 a reel	5–11c $ 10.00 to $ 50.00 a reel	20–25c	4–20c	3–30c	13–30c
United States	5–12c	7–15c $ 8.40 to $ 12.40 per reel	12–25c	3–10c		
British	3–10c	2 to 4 1/2c	25c	3–15c		
Inter. Orgs.			20–62c			
European	3–22c	2–13c		6–60c	10c to $ 1.72	13c to $ 1.50
Hispanic	15c		$ 1.00	10–18c	10c	

It is to be noted from this table that many of the highest cost figures are, not surprisingly, in the developing areas. Institutions in these areas do not, however, use priced exchange to recoup their costs rather than some type of piece-for-piece exchange any more than institutions in the other areas (see Question 6, below).

6. *On what basis are your photoreproduction exchange transactions conducted?*

	Page-for-page or vol.-for-vol. of copied material	Title-for-title	Frame-for-frame	Priced exchange	Other	No reply
African-Asian	11	5	0	4	9	1
United States	3	2	0	7	5	0
British	1	4	2	2	8	1
Inter. Orgs.	0	1	0	1	3	0
European	6	5	3	8	16	1
Hispanic	2	2	0	0	0	0
Totals	23	19	5	22	41	3

Most of the forty-one who replied "other," said they used a combination of these bases, a reel-for-reel basis, or sent all materials on "open" exchange.

The reason for the world-wide unpopularity of priced exchange vis-à-vis other bases of exchange for reprographics, is of course, not difficult to discern. Priced exchanges normally require a great deal of bookkeeping and accounts are constantly affected by the ever-fluctuating international monetary exchange rates. Owing to these changes the monetary value of materials offered in exchange cannot readily be recouped and the time involved in acquiring the material requested in exchange is usually great enough to witness still further, and often unfavorable fluctuation on the exchange rates, while the staff overhead continues high. In reprographic work, where all of these factors are accentuated, page-for-page, title-for-title, frame-for-frame, or any combination of these exchange bases, seem more realistic than priced exchange.

7. *Do you supply exchange partners with some photoreproductions free-of-charge if these materials do not exceed a certain limit?*

	Yes	No	No Reply
African-Asian	17	13	0
United States	7	9	1
British	10	7	1
Inter. Orgs.	4	1	0
European	27	7	5
Hispanic	2	1	1
Totals	67	38	8

It is interesting to note that the practice of sending some photoreproductions free is most common in the European area, and is followed by a majority of the institutions in all areas, except the United States. Following is a cross-section of the limits set by the sixty-seven institutions which supply some microforms free-of-charge within certain limits:

African-Asian – Six institutions said they had no firm limits or the limits were discretional. Money limits varied from $ 1.00 to $ 5.00, page limits from five to twenty.

United States – Many said "no firm limits", or "reasonable limits," or "a couple of pages". Stated page limits varied from ten to twenty.

British – Page limits varied from ten to one-hundred, dollar limits from $ 2.00 to $ 10.00. Some commented that the limits varied with the volume of the exchange.

International Organizations – Page limits from "a few pages" up to thirty pages. One organization said the limit was one report.

European – Limits were from five to fifty pages, from ten to seventy frames or xeroxes, and from $ 2.00 to twenty marks ($ 8.80).

Hispanic – From two to three xeroxes; one set a dollar limit of $ 5.00.

8. *Is your program for the exchange of reprographic materials coordinated with your library's preservation program?*

	Yes	No	No Reply
African-Asian	8	20	2
United States	4	13	0
British	3	13	2
Inter. Orgs.	0	4	1
European	13	24	2
Hispanic	2	2	0
Totals	30	76	7

This question was derived from experience in the Library of Congress, where material to be filmed for other organizations is automatically considered for filming in entirety for preservation purposes. The predominant comment in nearly all the thirty affirmative replies, however, referred to the practice of filming various types of books and newspapers for preservation purposes, and then making positive film copies available to their exchange partners.

9. *Do you find that copyright or other restrictions (e.g. no copying allowed of rare or brittle materials) on reproduction of materials constitute a significant problem in your reprographic exchange program?*

	Yes	No	No Reply
African-Asian	8	21	1
United States	4	13	0
British	0	16	2
Inter. Orgs.	0	4	1
European	12	26	1
Hispanic	1	3	0
Totals	25	83	5

Only 25 respondents to Question 9 indicated that copyright or other restrictions on reproduction of materials constituted a significant problem in their reprographic exchange programs. They noted such difficulties as: (1) having to get written permission to copy slows down the exchange, (2) it is not easy to get in touch with authors to obtain permission to reproduce, and (3) must exclude rare and brittle materials to avoid damage to them. Some indicated they will copy only uncopyrighted materials or their own publications. One U.S. university said it was "impossible to send reproductions of out-of-print American titles to exchange partners".

10. *In your experience, approximately how long does it take the laboratory to complete the average order for photoreproductions for exchange use?*

	Several days	Several weeks	Several months	Various	No Reply
African-Asien	13	13	2	0	2
United States	12	2	1	1	1
British	9	4	0	0	5
Inter. Orgs.	4	0	0	0	1
European	13	12	1	11	2
Hispanic	3	0	0	1	0
Totals	54	31	4	13	11

Here again is another question which is undoubtedly closely related to Questions 2A and 2B, which showed a preference for the use of xeroxes in exchange. Xeroxes can be made quickly and with a minimum of fuss and paperwork. Any of the reprographic products which take longer to prepare will require additional record-keeping to route the material, when finally received, to the proper exchange partner.

11A. *Does the laboratory which produces your photoreproductions follow certain technical specifications in their manufacture (i.e. with regard to film base, washing of film to eliminate residual hypo, length of leaders, etc.)?*

	Yes	No	No Reply
African-Asian	13	10	7
United States	10	4	3
British	7	2	9
Inter. Orgs.	2	1	2
European	18	15	6
Hispanic	0	2	2
Totals	50	34	29

11B. *Can the laboratory meet the special requests of some exchange partners for photoreproductions produced to the technical specifications they wish to have?*

	Yes	No	Sometimes	No Reply
African-Asian	10	13	1	6
United States	6	6	3	2
British	7	4	0	7
Inter. Orgs.	1	2	0	2
European	15	17	1	6
Hispanic	0	3	0	1
Totals	39	45	5	24

Of the eighty-four replies received to the Question 11A, fifty libraries, or nearly sixty percent, stated that their laboratories did follow certain technical specifications. Many mentioned International Standards Organization standards, while others referred to the specifications of the National Microfilm Association, the American National Standards Institute, the British Standards Institute, the Library of Congress, other national standards institutions, or to the standards of various manufacturers of photographic equipment. There were a few references to "ordinary technical specifications," "international standards," and "state standards" – all unspecified.

To the last question, 11B, thirty-nine out of eighty-nine replies indicated that laboratories could meet special requests for materials produced to certain technical specifications, forty-five said no, and five said "sometimes," with the British area showing the greatest degree of flexibility (seven out of eleven will respond to special requests).

Conclusion

Is it possible to summarize the findings of our inquiry and try to draw general inferences about the use of reprographic materials in the international exchange of publications? We can probably safely assume that most of the 290 institutions that did not reply at all to the questionnaire do not use microforms in exchange to any significant extent. If this assumption is valid, then only about one-sixth of the institutions queried *do* use reprography in exchange. Slightly more than ten percent of the institutions queried regard reprographic materials as an important part of their exchange program. Less than one percent of the institutions queried use more than ten percent of their exchange funds to acquire microforms. On the other hand, the use of these materials in exchange has been steadily increasing since 1950, and the rate of increase has more than doubled since 1960. Although all types of photoreproductions are used, microfilm and xerox are clearly predominant. Little uniformity appears to exist in the use of technical specifications and in the exchange bases used by the respondents, and these two areas may perhaps present opportunities for further explorations.

* Chief, Exchange and Gift Division, Library of Congress, Washington

A Recommended Format for International Exchange Lists of Publications

A. Allardyce*

Exchange of publications is often based on bibliographies and catalogues produced for other more general purposes. However in many cases, and normally for duplicates, special lists of material available for exchange have to be compiled. In these cases uniform presentation is an advantage. The IFLA Section on the Exchange of Publications accepted a suggestion in 1975 that a standard format should be sought and as a result the following was produced. The original draft benefited from the Section Committee's suggestions.

RECOMMENDED FORMAT FOR INTERNATIONAL EXCHANGE LISTS OF PUBLICATIONS:

1. Each list should have a distinctive number e.g. a serial number.

1.1 This number should be clearly displayed on each page.

2. Each item should have a distinctive number.

2.1 If these numbers are not in one series, each section should be numbered so that list, section and item numbers together identify an item.

3. Each list should have a descriptive title e.g. of the subject.

PRESENTATION RECOMMANDEE POUR LES LISTES D'ECHANGES INTERNATIONAUX DE PUBLICATIONS:

1. Chaque liste doit avoir un numéro distinct, par exemple un numéro de série.

1.1 Ce numéro doit être clairement mis en vedette sur chaque page.

2. Chaque publication proposée doit avoir également un numéro distinct.

2.1 Si l'ensemble des numéros ne forme pas une seule série, chaque section doit être numérotée de façon que les numéros de liste, de section et d'article permettent ensemble d'identifier une publication.

3. Chaque liste doit porter un titre qui la décrive, par exemple indiquant le sujet des publications.

4. The method of allocation (e.g. on a system of priorities or first come first served), and if applicable a closing date should be given, to enable libraries to judge the time available for checking.

5. The recommendations, where relevant, also apply to cards offering material.

4. On doit indiquer la méthode d'attribution (qu'il s'agisse par exemple d'un système de priorités, ou que les premières réponses soient les premières servies) et si possible une date limite, afin de permettre aux bibliothèques de juger le temps dont elles disposent pour le pointage de la liste.

5. Les recommendations s'appliquent également, le cas échéant, aux offres qui seraient inscrites sur des fiches.

* Head, Monograph Acquisitions and Gift and Exchange, British Library Lending Division, Boston Spa, Wetherby, West Yorks

A Costing Analysis of the Gift and Exchange Section of the British Library Lending Division (based on an internal report by R. Steemson)

E. Tatterton, K. King and A. Allardyce*

Introduction

The Gift and Exchange Section (GES) of the British Library Lending Division emerged from the amalgamation of the British National Book Centre of the National Central Library and the Donations Section of the National Lending Library for Science and Technology. These libraries joined together in 1973 to form the British Library Lending Division. The function of GES is to receive suitable surplus material from British libraries and to place it if possible in the national collections and in other libraries both at home and abroad.

GES accepts all kinds of printed library materials and also offers material kept available for disposal by other libraries. Items are first checked against the loan stock of the Division and material not required is offered on lists to British and then to foreign libraries.

Although approximate costs of the operation have always been known to the management, a better estimate of the direct use of staff on specific types of material was attempted in 1977, based on timed sample tests.

The IFLA Section on the Exchange of Publications thought that some analysis of the economics of duplicate exchange would be useful to those operating or considering operating duplicate clearing houses.

Consequently, the findings were distributed to the 1977 meeting of the IFLA Section on the Exchange of Publications in a paper by R. Steemson. The present paper is based on more recent work carried out by Mr. Steemson. There is a condensed version of it[1] and it was also originally prepared in slightly shorter form for the 1978 IFLA 44th Council Meeting.

Preliminary notes

This was a team exercise. A diary programme was designed and carried out by R. Steemson, who also wrote the bulky internal report on it. E. Tatterton, at that time the General Manager of Monograph Acquisitions and Gift and

Exchange and K. King the Supervisor of GES checked, selected and compressed the data under the direction of A. Allardyce, who edited the results.

The object is to show a scale of approximate direct costs for every significant part of the GES operation. To do this a record was kept during May 1978 by each person or team of the time spent and the work accomplished in connection with each identifiable GES process. In addition sample surveys were used to establish certain approximate equivalents. Greater accuracy would not have been useful, or possible without much more staff time than was available.

The main stages of each operation were first flow-charted and block diagrams were constructed (Sections 1 and 2/(p. 79–80). Then tables of the operations, numbered, were made (Section 3/(p. 82–83). Diary forms were issued to staff or team leaders, after the trade unions had been informed that the object was to collect management information and not to measure the performance of individuals.

Only those parts of the operations on which sufficiently accurate management information did not already exist were fully recorded.

Tables 1/(p. 85) and 2/(p. 86) show the aggregate time spent on each operation (assembled from the diaries) and the amount of work done (extracted from the diaries or from other information about the task).

Since salaries are on Civil Service scales, the cost of the time was then calculated by reference to Civil Service Department staff cost tables.

Exclusions: Accommodation, heating, lighting, cleaning, administration (including senior management costs and middle management unless directly participating), common services except where directly involved, carriage.

The unit costs of each operation in a sequence were then calculated in terms of the significant unit easiest to identify. Thus Section 3 shows costs per book, entry, issue, request. In tables 5/(p. 91) and 6/(p. 92) entries and requests have been reduced to books and periodical pieces after a separate small sampling exercise.

The GES operation at the Lending Division is a byproduct to some extent of the acquisition process, and some of the work which is necessary for the former has already been done for the latter. Table 5 therefore shows as far as is practicable two costs – a marginal cost and a "total" cost. The marginal cost is the significant one in the British Library but the total cost would be that incurred in a dedicated exchange centre.

Tables 5 and 6 should be distinguished. The cost per item handled at each stage is given in Table 5 and this is an important part of the raw material for making decisions about staff and methods. Table 6 on the other hand, shows the cost per item successfully placed during a certain period. Normally this cost would not be derived from these restricted data since it is only long term figures which are significant and they should relate to the overall cost. Nevertheless the present derived costs are given for interest.

No universal value is claimed for these figures. The nature of the material, the quality of staff and the degree of interest of requesting libraries vary enough from time to time and from place to place to make such a claim excessive. But together with other such studies as take place, this study should be some guide to the *order* of costs incurred in accomplishing a given result.

The arrangement of the following sections is:

1. Organization of GES
2. Operations and block diagrams
3. Diary exercise
4. Unit costs extracted from 3

Appendix – Rapid listing process for batch cataloguing of monographs, by M. Barwick and A. Allardyce

1. *Organization of GES*

GES has a supervisor at second professional level who reports to the General Manager of Monograph Acquisitions. He in turn reports to the Head of Monograph Acquisitions and Gift and Exchange who is responsible for policy. GES is organized in two units each with a manager at first professional level.

The GES Depot receives, handles and despatches material donated by other libraries or weeded from the Division's loan stock. In addition to the manager, there is a staff of 8 Library Assistants and 2 Clerical Assistants.

The GES Office receives offers on cards both from the Depot and from other libraries, prepares exchange lists and allocates materials requested. In addition to the manager there is a staff of 5 Library Assistants.

Monograph cataloguing is carried out in Monograph Acquisitions by 1 first professional level member of staff and 2 Library Assistants. Typing and production of the lists by photolithography is carried out using the Library's centralized facilities.

2. *Operations and block diagrams (see pages 79 to 83)*

The block diagrams represent the processes as carried out at the time of the diary exercise. Some methods may have been dictated by temporary factors such as the present lack of accommodation at the Lending Division and therefore may not reflect the most efficient approach.

GES Depot

Figure 2.1.a gives details of the sorting of incoming donations together with the handling procedures for periodicals.

For the purpose of this exercise a periodical "run" is defined as at least 3 consecutive issues or a substantial number of odd issues of the same title. As the Division's collection of periodicals is fairly comprehensive from 1960 onwards, only pre-1960 periodical issues and "runs" are checked against the loan stock. At present there is no comprehensive record of the Division's periodical stock, therefore checking is done at the shelf.

In a typical year one mile of periodicals is received of which 95 % becomes available to other libraries.

If the rate on donated Monographs exceeds the cataloguing capacity of the section the excess is temporarily housed in a buffer store 1 1/2 miles away. Cataloguing is by a "rapid listing" batch system described in the Appendix. Monographs are added to stock as necessary and the remainder made available to other libraries through the Office. Before offering, a three figure Dewey class mark is given to the book so that the lists can be arranged in broad subject areas.

In a typical year, approximately 45,000 monographs are received of which 40 % become available for offer.

GES OFFICE

HANDLING OF INCOMING DONATIONS & PROCESSING PERIODICALS

Fig. 2.1.a

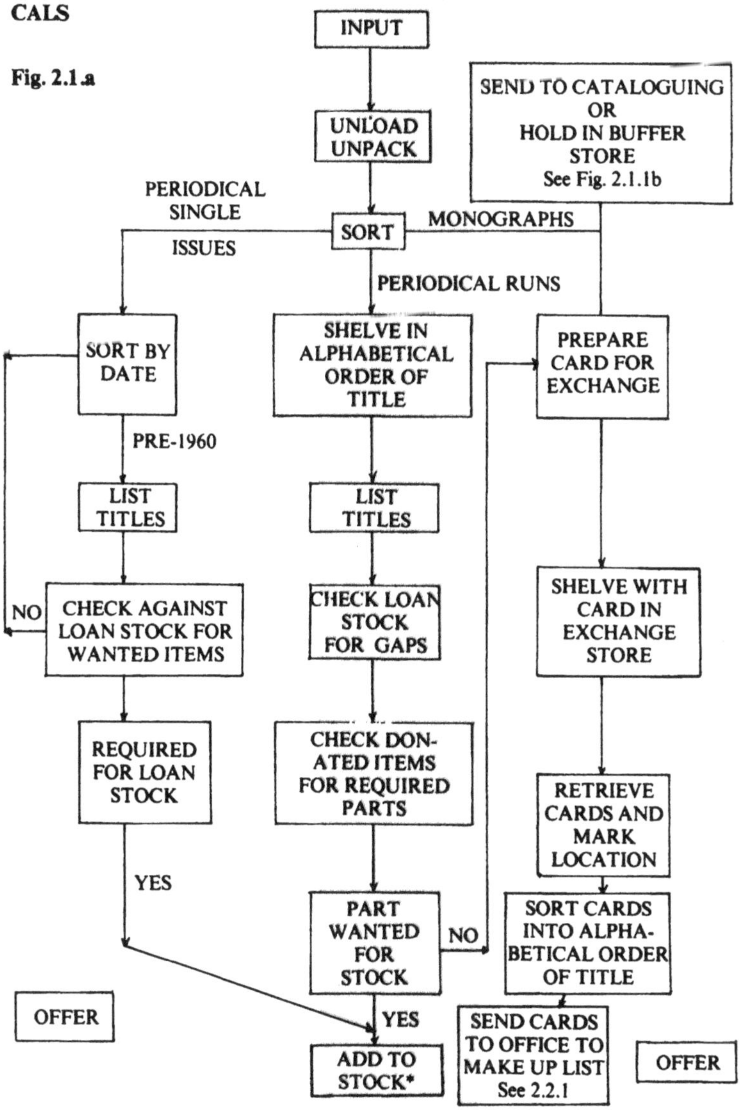

* At present stored until shelving available

GES DEPOT

PROCESSING DONATED MONOGRAPHS

Fig. 2.1.b

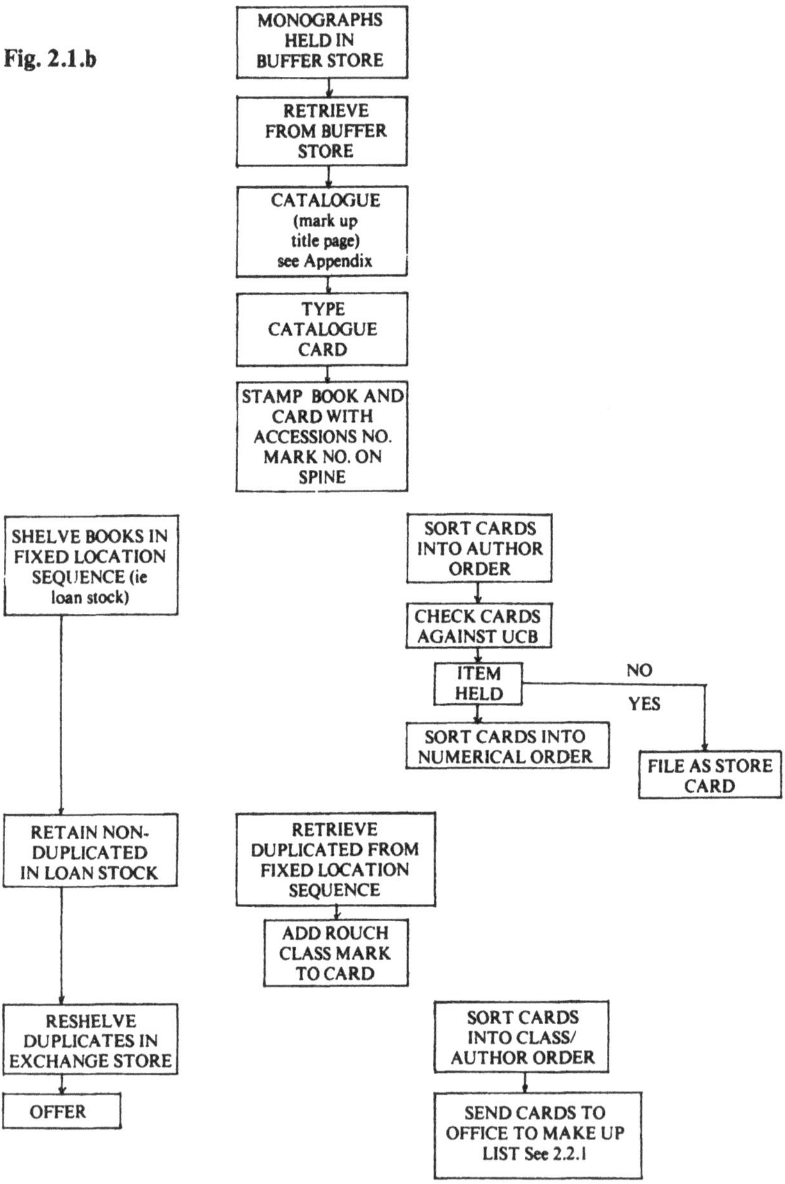

GES Office

Figure 2.2.1 deals with the preparation and despatch of the UK lists – GES Books and GES Periodicals. These include both material in the GES Depot (internal offer) and that offered on cards by UK Libraries to the GES Office (external offer). Before listing external offers are checked against the catalogue and those required are requested at once. External offers are eventually despatched by the offering library direct to the requesting library.

Material unallocated from GES Books and GES Periodicals is then offered on GES International. All lists quote a closing date and otherwise conform to the IFLA Section on the Exchange of Publications Recommended Format. [2] The total number of titles offered on GES International last year was 25,000.

Allocation

Figure 2.2.2 deals with the allocation of GES Books and GES Periodicals. Allocation takes account of the need to

1. reinforce special subject collections,
2. accommodate libraries requesting relatively few items,
3. avoid splitting useful runs.

For GES International there are less material and fewer libraries and therefore allocation is much simpler.

3. *The Diary Exercise*

In this exercise, each task or group of associated tasks was given a code number as shown in tables 1 and 2. In drawing up the list of tasks the following points were considered:

a) Where a sequence of distinct operations was by one person working through a batch of material, the sequence was taken as one task. The task list thus reflects more the way the work flow is organized than the actual individual steps undertaken.

b) A measure of the output was required in some cases. Whenever possible this was estimated from routine statistics collected, or on the principle that where one task follows directly from another, the total output of the second task must equal the total output of the first. In some cases estimates had to be made for the diary exercises.

GES OFFICE
PREPARATION OF GES BOOKS AND GES PERIODICALS

Figure 2.2.1

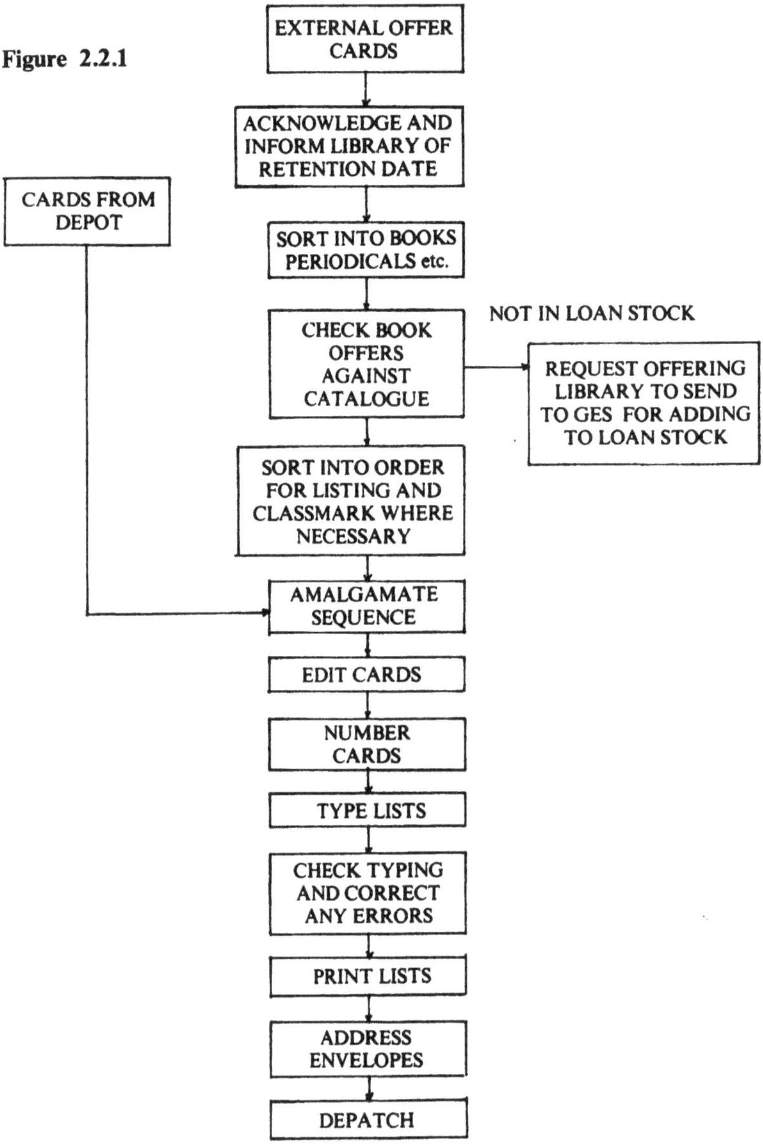

See Figure 2.2.2 for Allocation procedures

GES OFFICE

ALLOCATION FROM GES BOOKS & GES PERIODICALS

Figure 2.2.2

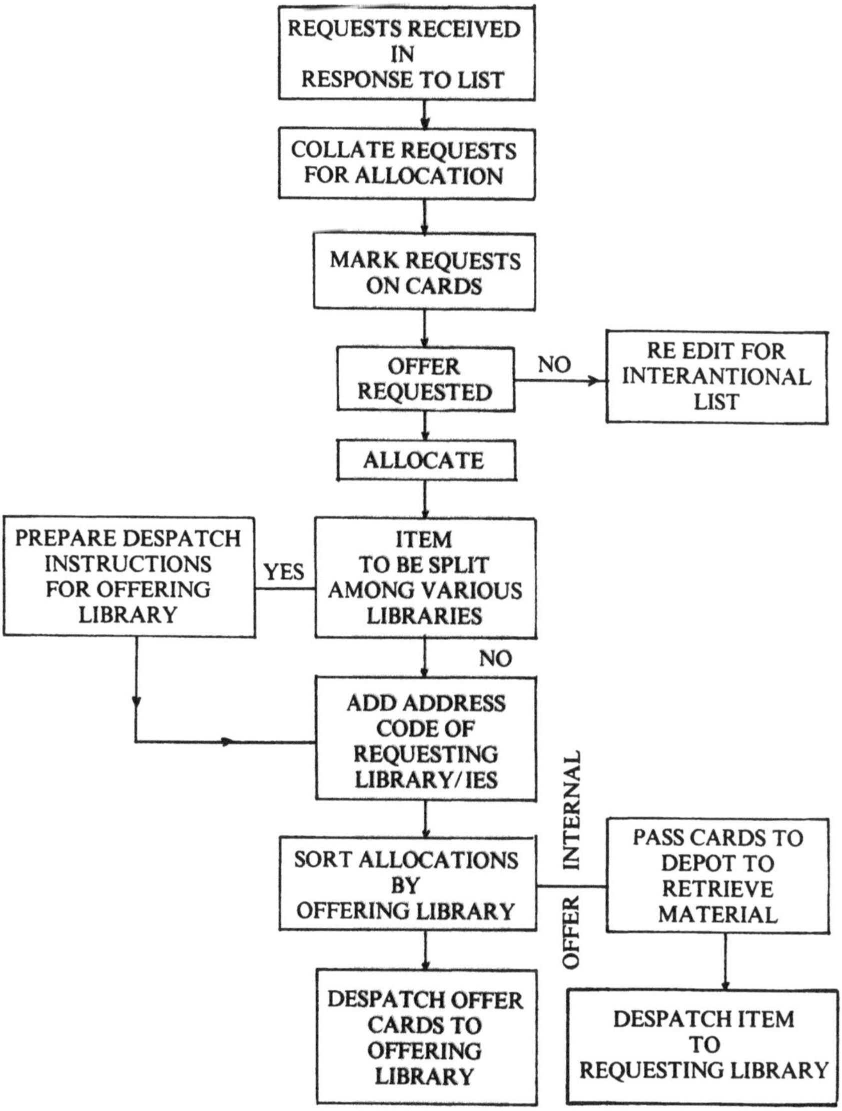

The time shown against each code is the total time of all grades of staff employed on that task. Simple division will not therefore produce the unit cost. Each block of time will have its own average value depending on the proportion given by the separate staff grades, and these complex calculations are omitted from this summary.

The Basic Staff Cost per hour of each grade was taken from the Civil Service Ready Reckoner of Staff Costs for 1977. The Basic Staff Cost is the average salary cost plus £ 700 per person to cover the average annual use of stationery, telephone, postage and small office machines. The hourly rate is calculated by dividing the total by 1650 hours (223 working days at 7.4 hrs). This takes into account leave, public holidays and sickness.

	£
Basic clerical grade	1.68 per hour
Typist	1.72 per hour
Library Assistant	2.05 per hour
1st Professional level	3.14 per hour

TABLE 1 – DEPOT TASKS MONITORED IN THE DIARY EXERCISE

CODE	TASK	TIME	OUTPUT
D10	Unload, unpack and primary sort	68 hrs	3,600 books 330 periodical titles (approx. 100 feet)
D20	Check against stock – periodical runs	28 hrs	190 titles (approx. 94 feet)
	Check against stock – periodical single issues	11 hrs	140 titles (approx. 6 feet)
D30	Card for offer – periodical runs	152 hrs	700 titles (approx. 580 feet)
	Card for offer – periodical single issues	20 hrs	590 titles (approx. 27 feet)
D40	Remove periodical cards for offer	3 1/2 hrs	700 titles
D51	Sort book cards for offer	24 hrs	2,913 cards
D52	Sort book cards for filing in catalogue	6 1/2 hrs	490 cards
D53	Sort cards for retrieval from stock	6 hrs	1,200 cards
D60	Retrieve duplicate books from stock	25 hrs	1,200 books
D61	Retrieve periodicals for despatch	48 hrs	6,270 pieces
D62	Retrieve books for despatch	44 hrs	1,758 books
D70	Pack, weigh, secure, invoice and despatch	152 hrs	7,150 periodical pieces 3,365 books
D80	Transfer of monographs to and from buffer store	78 hrs	-

TABLE 2 – OFFICE TASKS MONITORED IN THE DIARY EXERCISE

CODE	TASK	TIME	OUTPUT
010	Sort external offers into books and serials	3 1/2 hrs	2,400 cards
011	Check external offers against stock (books only)	8 hrs	366 books
020	Sort offer cards into order for listing including allocation of class marks where necessary (i) Books	2 hrs	2,980 cards
	(ii) Periodicals	30 hrs	4,250 titles
021	Edit (i) GES Books	27 hrs	2.980 books
	(i) GES Periodicals	63 hrs	4,250 titles
022	Re-edit for GES Internatio-nal	21 hrs	3,190 entries
023	Number cards	4 hrs	3,190 cards
030	Proof read lists	32 hrs	5,020 entries
040	Annotate and collate requests	43 hrs	8,900 book requests 1,512 periodical requests
050	Sort offer cards for allocation	17 hrs	4,220 book requests 1,512 periodical requests
	Allocation:		
051	Internal offers – single re-quests	21 hrs	980 requests
052	Internal offers – multiple requests	6 hrs	103 requests
053	External offers – single requests	9 1/2 hrs	320 requests
054	External offers – multiple requests	18 hrs	N/A

055	Allocation of requests for GES International	27 1/2 hrs	970 requests
080	Address allocations	2 hrs	1,600 requests (approx. 85 libraries)

Sufficient data is available to enable the following operations (Coded X) to be costed without a diary exercise:

Cataloguing, typing catalogue card, stamping number on monograph, filing catalogue card, typing and printing lists, despatch of lists, notifications.

Unit costs extracted from 3. The diary exercise

Table 3 – Monographs

1. Cost of adding a book to stock excluding shelving

Code	Internal Offer		Cost /Unit
D10	Unpack	2.0	
X	Catalogue and Process	12.2	
X	Check and file Catalogue card	8.3	22.5 pence/book
	This cost is approximately twice that for filing catalogue cards, since the 50 % books wanted for loan stock are only identified at this stage.		
	Note A cost of 5p is at present incurred by the temporary housing of material in a Buffer Store. This has not been included in our calculations.		
	External Offer		
	As Above	22.5	
011	Check carded offers in UCB	4.6	27.1 pence/book

	2. Preparation of GES Books		
	Internal Offer		
D51	Sort cards for offer	4	
D53	Sort cards for retrieval	1	
D60	Retrieve Books	4	
021	Edit Cards	2	
023	Number Cards	0.2	
030	Proof Read List	1	
X	Type and Print List	10	
X	Despatch and Postage	4	27.2 pence/entry
	External Offer		
020	Sort Cards for Offer and notify retention date	6	
021	Edit Cards	2	
023	Number Cards	0.2	
030	Proof Read List	1	
X	Type and Print	10	
X	Despatch	4	23.2 pence/entry
	3. Allocation from GES Books		
	Internal Offer		
040	Record Requests and Annotate Cards	2	
051–054	Allocation	8	
D62 & D70	Retrieve Books and Despatch	10	20 pence/book
	External Offer		
040	Record Requests and		
051–054	Annotate Cards	2	
X	Allocation	8	
	Notification	3	13 pence/book
	4. Preparation of GES International		
022	Re-edit List	3	
023	Number Cards	0.2	
030	Proof Read List	2	
X	Type and Print	9	
X	Despatch and Posting	7	21.2 pence/entry

5. Allocation from GES International

	Internal Offer		
055	Allocation	6	
D60 & D70	Retrieve Books and Despatch	10	16 pence/book
	External Offer		
055	Allocation	6	
X	Notification	7	13 pence/book

Table 4 – Periodicals

1. Cost of adding periodicals to stock excluding shelving costs

CODE	Runs		Cost /Unit
D10	Unload, unpack etc	2	
D20	Check against loan stock	14	16 pence/issue
	Odd Issues		
D10	Unload etc	2	
D20	Check against loan stock	71	73 pence/issue

2. Preparation of GES Periodicals

	Internal Offer		
D30	Card Material	47	
D40	Remove Cards	1	
D51	Sort for Offer	3	
021	Edit Cards	4	
023	Number Cards	0.2	
030	Proof Read List	2	
X	Type and Print	12	
X	Despatch and Postage	5	74.2 pence/entry
	External Offer		
020	Sort for offer	3	
021	Edit cards	4	
023	Number cards	0.2	
030	Proof read list	2	
X	Type and Print	12	
X	Despatch and postage	5	26.2 pence/entry

	3. Allocation of GES Periodicals		
	International Offer		
040	Record and Annotate requests on cards	5	
051–054	Allocate	23	
D61	Retrieve Material	5.5	
D70 & 080	Despatch	8.5	42 pence/request
	External Offer		
040	Record and Annotate requests on cards	5	
051–054	Allocate	23	
X	Notification	3	31 pence/request
	4. Preparation of GES International		
022	Re-edit list	3	
023	Number Cards	0.2	
030	Proof Read	2	
X	Type and Print	9	
X	Despatch and Postage	7	21.2 pence/entry
	5. Allocation from GES International		
	Internal Offer		
040	Record and annotate requests	5	
055	Allocate	6	
D61	Retrieve Material	8	
D70 & 080	Despatch	7.5	26.5 pence/request
	External Offer		
040	Record and annotate request	5	
055	Allocate	6	
X	Notification	3	14 pence/request

Table 5 – Cost of handling material (in pence per item)

	Marginal (ie residual to acquisitions)		Total (ie as if a self-contained operation)	
	Internal offer	External offer	Internal	External
GES Books				
Unpack	–	–	2	–
Catalogue etc	–	–	12.2	–
List	22.2	23.2	22.2	23.2
Allocate and despatch	20	13	20	13
Total cost for UK distribution	42.2	36.2	56.4	36.2
GES Periodicals (12 pieces per title requested, 13 pieces per title listed)				
Unpack	–	–	2	–
Sort and list (piece listed)	5.7	2	5.7	2
Allocate and despatch (piece requested)	3.5	2.6	3.5	2.6
Total cost for UK circulation	9.2	4.6	11.2	4.6
GES International				
Books				
List	21.2	21.2	–	–
Allocate and despatch	16.0	13.0	–	–
TOTAL	37.2	34.2	–	–
Periodicals (5 pieces per title listed, 10 pieces per title requested)				
List (piece listed)	4.2	4.2	–	–
Allocate and despatch (piece requested)	2.6	1.4	–	–
TOTAL	6.8	5.6	–	–

Table 6 – Marginal cost of redistribution (in pence per item)

	Internal offer	External offer
GES Books Approximately 60% take-up		
List	37.0	38.7
Allocate and despatch	20.0	13.0
TOTAL	57.0	51.7
GES Periodicals Approximately 35 % take-up, 13 pieces per periodical title listed and 12 pieces per periodical title requested		
Unpack/sort	–	–
List	16.3	5.7
Allocate and despatch	3.5	2.6
TOTAL	19.8	8.3
GES International		
Books – 26 % take-up		
List	81.5	81.5
Allocate and despatch	16.0	13.0
TOTAL	97.5	94.5
Periodicals – 30 % take-up; 5 pieces per title listed; 10 pieces per title requested		
List (piece listed)	14.0	14.0
Allocate and despatch (piece requested)	2.6	1.4
TOTAL	16.6	15.4

The Cataloguing of Donated Monographs at the British Library Lending Division

M. Barwick and A. Allardyce

The Division uses a batch-cataloguing system called „Rapid Listing“, (which was first developed by A. Allardyce at the National Central Library in 1962) in order to reduce the cataloguer's effort on each book and cut cataloguing costs by streaming the material by levels of cataloguing difficulty.

It is possible to use this system, because only brief cataloguing records are required for entries in the Union Catalogue of Books (UCB), which is filed on the Berghoeffer system (by author's surname and book title), and for the production of the Gift and Exchange lists in which residual titles are offered.

The batch-cataloguing system is based on the principle that non-professional staff can be trained to identify, by a glance at the title page, a limited range of entries they can construct at once and to pass the rest on to a professional cataloguer. The system also relieves them of the physical work of typing (except for cross-references, which are kept to a bare minimum) and cuts writing down to mere coding.

The primary batch varies with the experience of the non-professional but covers mainly English language monographs, excluding official publications, monographs in series, biblical works, reference books and exhibition catalogues. These items, together with those foreign language works which are outside the linguistic range of the non-professionals (some of whom are graduates), form the secondary batch, with which the professional cataloguer deals. The latter is also responsible for training and supervision.

After an initial training period, the average non-professional member of staff can scrutinize books and catalogue approximately sixty of them per hour although as many as a hundred an hour have been achieved regularly by experienced staff. The system was designed on the assumption that this rate can be maintained only for a relatively short period, and that thereafter production would lag. In order to maintain momentum therefore staff catalogue for only four hours a day (with a short break) and perform more relaxed tasks for the remainder.

Although the proportion of material which can be dealt with at this rate varies from consignment to consignment of donations, often as much as 90 % is suitable for rapid listing. Public library donations, for example, are usually quite straightforward, whilst often many items donated by special and university libraries require the cataloguing expertise of the professional member of staff.

The cataloguing rules used are based on AACR 1967, modified to the practice of the UCB. Non-professional staff may briefly refer to the rules for confirmation, but they are encouraged to catalogue only those items for which they can immediately recognize the appropriate heading and form of entry. Cataloguing problems which require more than a basic knowledge of the rules are passed on straight away. The following AACR rules indicate some of the exclusions:

5B	Collections without collective titles.
7	Adaptor or original author.
12	Praeses or respondent.
17C	Works by head of government.
19	Related works.
20–26	Laws, court decisions etc.
27–32	Religious publications.
46E–51	Nobility, royalty, saints, Roman names, etc.
54-58	Arabic, Thai names.
92–92	Religious bodies and officials.
108–119	Bible and other scriptures.

The catalogue entry is produced by marking up the title page very explicitly (for the sake of a succession of typists from a typing pool). The following code numbers are written in pencil above the first word of the appropriate field:

1. Author's surname or corporate body.
2. Author's forename or subordinate body.
3. Title.
4. Edition statement.
5. Place of publication.
6. Publisher.
7. Date of publication.
8. Series note.
9. ISBN.
10. Any other details e.g. "2 copies".

Typists then type catalogue cards using the marked-up title page as a basis. Cards are typed in this way at an average rate of forty per hour, although higher rates have been regularly achieved.

A typical marked-up title page and the resultant card are shown in figures 1 and 2.

To keep unit costs to a minimum, only spot checks are made on both the cataloguers' cards and typists' output. However, anomalies in catalogue entries are usually identified at the filing stage and these prove relatively few.

Clerical staff give accession numbers to the books on the title-page and spine and to the cards at a rate of approximately forty per hour. The completed cards are alphabetised, and then checked against UCB at a rate of approximately

(OXFORD MEDICAL PUBLICATIONS)

A PRACTICAL MANUAL OF DISEASES OF THE CHEST

BY

MAURICE DAVIDSON

M.A., M.D. OXON., F.R.C.P. LOND.

PHYSICIAN TO THE BROMPTON HOSPITAL FOR CONSUMPTION AND DISEASES OF THE CHEST (SOMETIME DEAN OF THE BROMPTON HOSPITAL MEDICAL SCHOOL); ASSOCIATE PHYSICIAN TO THE ROYAL CANCER HOSPITAL; CONSULTING PHYSICIAN TO THE MILLER GENERAL HOSPITAL FOR SOUTH-EAST LONDON; CONSULTING PHYSICIAN TO THE WESTERN OPHTHALMIC HOSPITAL; HON. CONSULTING PHYSICIAN TO THE PAPWORTH VILLAGE SETTLEMENT, CAMBRIDGE

THIRD EDITION

GEOFFREY CUMBERLEGE
OXFORD UNIVERSITY PRESS
LONDON NEW YORK TORONTO
1948

Fig. 1

Figure 2

Davidson, Maurice
A practical manual of diseases of the chest.
3rd ed.
London, Oxford University Press, 1948.
(Oxford medical publications)

eighty per hour. About half are filed as new entries for stock and the rest passed on to the Gift and Exchange Section for inclusion in their lists.

The overall unit cost of cataloguing, typing, processing and filing straightforward material is approximately 19p[3], excluding supervisory costs. More complex cataloguing performed by the professional member of staff may cost up to half as much again.

References

* Head, Monograph Acquisitions and Gift and Exchange, British Library Lending Division, Boston Spa, Wetherby, West Yorks

1 A. Allardyce. Some Operational Costs of the Gift and Exchange Section of the British Library Lending Division. Interlending Review 7 (3) 1979, p. 90–98.

2 See A. Allardyce. A Recommended Format for International Exchange Lists of Publications (See p. 73–74).

3 Figure in January 1978.

The Economic Aspect of the Exchange of Duplicates. Time Studies on Books. A Case Study

Rita Ejlersen*

In continuation of the analysis carried out by the Gift and Exchange Section of the British Library [1], and of USBE [2] the IDE performed time studies on its exchange of duplicate monographs over a 10-month period.

It is hoped that the present study may provide supplementary information on the exchange of duplicates to institutions performing or intending to establish such exchanges. With a view to this, the paper gives a detailed description of working methods which have been found time-saving at the IDE.

In many respects the working conditions of IDE differ from those of the above institutions. Not even the results of the analysis of the exchange of duplicate monographs in the British Library and IDE are comparable: IDE's duplicates are used exclusively for exchange purposes, domestic as well as international. All duplicates are received at the Institute and are catalogued there. Solely duplicates intended for exchange are catalogued, and only one set of lists is prepared and distributed among domestic and foreign libraries simultaneously.

The material examined at IDE is less extensive than that of the other institutions. The analysis of the processes is expressed in net working time excluding rest periods. The figures were obtained over periods of consecutive hours or days.

The examination was limited to time studies, while no cost studies were undertaken for this paper. Cost studies would not be comparable, due to varying salaries of the different countries. Even within one's own country cost studies may vary considerably from year to year. The experience of recent years has shown how salaries may rise due to the inflation. But if required, cost studies may easily be deduced from the time studies.

Organization of the IDE

IDE, Institut danois des Echanges internationaux de Publications scientifiques et littéraires, is a Government institution. It was founded in 1946 with the aim of coordinating and centralizing Danish exchanges. The Institute is in charge of the exchange of official publications with 14 countries and of the exchange of doctoral theses and other publications issued by the University of Copenhagen.

The Institute edits two annual bibliographies which are based on the material received for the above two exchanges: Impressa publica Regni danici (Bibliografi over Danmarks offentlige publikationer), which lists Danish official publications and all Danish dissertations, and a yearly bibliography of the doctoral theses and other publications of the University of Copenhagen.

IDE handles the exchange of periodicals of the Danish National Library (Det kongelige Bibliotek), and of the State and University Library of Århus (Statsbiblioteket i Århus). Further IDE runs a dispatch service for Danish Government institutions and, last but not least, the Danish exchange of duplicates is centralized in IDE.

The staff consists of 9 members: two academic librarians, one non-academic librarian, one staff member previously experienced as a book-seller, clerical employees, and a half-day driver. Besides, various trainees work at the Institute for shorter periods of time.

Exchange of Duplicates

In my opinion centralization of the exchange of duplicates has many advantages.

At a centre the exchange of duplicates is an organized routine and not a casual job taken up now and then when time permits; therefore the duplicates of a centre may be made available a short time after they have been weeded out from the libraries.

Requests for missing material may be directed to one place and they have a greater chance of being complied with, since a comprehensive collection of duplicates will naturally offer a wider choice of publications and longer runs of periodicals.

Finally, the staff of a centre will acquire considerable experience in evaluating duplicates as well as an all-round knowledge of the needs and requirements of the individual partners at home and abroad.

Such experience also proves to be of advantage in the distribution of foreign offers. As mentioned, Denmark has a centralized system of duplicates exchange. All lists from abroad are sent to the IDE, which circulates the lists among relevant Danish research libraries. When the latter have made their selection, IDE undertakes the ordering of books requested by the libraries. Immediately on receipt the IDE distributes the material requested and keeps statistics of the number of publications received from each individual foreign partner.

Working Methods

The following table shows the various phases of treatment of a duplicate monograph, from its receipt at the Centre until it is forwarded to another institution. A table on p. 107–108 gives the results of a detailed analysis of each phase.

The average time employed is stated for each unit (except for units 4 and 11 requiring no calculable time, and excepting the printing of stencils which takes place outside the Institute), the number of books examined in each process, and the different categories of staff performing the work that was the object of time studies.

The work falls into three main parts: Acquisition, processing and distribution.

Acquisition

1. Unpack
2. Discard unwanted items
3. Sort and shelve received material in subject groups

Processing

4. Select duplicates for exchange lists
5. Catalogue books on slips and provide the books and slips with numbers
6. Arrange slips and books alphabetically
7. Proof-read slips
8. Type list from slips [printing of stencils]
9. Select recipients for list

Distribution

10. Dispatch lists to partners
11. Collect requests received
12. Mark requests on distribution list
13. Allocate duplicates
14. Keep statistics of outgoing duplicates
15. Pack duplicates and post them to partners

Staff in duplicates section

The member of the staff with many years' experience as a bookseller is in charge of the duplicates section and has no other fields of work. All other staff members (except one) work in the duplicates section when they can spare time from the various other exchange activities for which they are responsible.

Sources of duplicates

The majority of all large Danish research libraries, various other institutions and occasionally even private persons pass on their surplus material to the Exchange Institute, whose annual accession amounts to approximately 1,000 linear metres of books and periodicals (about 75,000 items).

The holdings are generally of the order of 900–1,200 metres, covering all subject fields. The duplicates are forwarded to IDE either in parcels or in containers; they may also be brought by car. Copenhagen libraries will frequently ask IDE to collect their duplicates.

Large consignments are generally announced beforehand by telephone. In order to avoid dispatch of superfluous material, such as ephemeral literature printed in a large number of copies, outdated scientific periodicals, popular textbooks and fiction, the Exchange Institute may ask libraries to roughly characterize the collections offered by them. In Copenhagen a staff member may visit libraries offering duplicates in order to select the publications which can be used for exchange purposes.

None of the books received are accompanied by catalogue slips; large consignments of periodicals occasionally include a list covering titles and years.

Acquisition

1, 2. *Unpacking of material and discarding of unwanted items*

A preliminary sorting of duplicates takes place simultaneously with the unpacking of them; generally 10–20 per cent prove to be of no value for exchange with research libraries and are discarded at once.

3. *Shelving in subject groups*

The material is sorted out and shelved in broad subject fields. Books and periodicals are shelved separately. Until 1976 the duplicates were split up into 24 groups, but it has been found that a consistent and reliable systematization in small groups requires qualified personnel to spend more time on the sorting out than we can well afford.

Accordingly, it was necessary to reduce the number of subject groups to 10, although we were aware that in many libraries lists covering narrower fields can be examined more conveniently.

So far we have found that the response to the duplicates lists has not declined after the introduction of a reduced number of subject groups.

Subject groups

I Humanities:

1. Literature, linguistics, music, theatre, film.
2. History, archaeology, art, architecture, politics, philosophy, religion, psychology, sociology, education, miscellanea.
3. Library science, general bibliography.
4. Economics, statistics, law.

II Medicine:

5. Medicine.
6. Pharmacology.
8. Odontology.
9. Medical abstracts and bibliography of medicine.

III Natural science:

9. Agriculture, biology, geography, meteorology, veterinary science, forestry, horticulture, fishery, botany, zoology.

IV Exact sciences:

10. Mathematics, physics, chemistry, engineering, astronomy, cybernetics, data processing.

Single issues of valuable periodicals are not included in the lists, but are shelved apart with a view to the possibility of establishing longer runs on the basis of future deliveries.

Processing

4. *Selection of duplicates for exchange lists*

The Institute endeavours to maintain a rotation of the various fields; but the choice naturally also depends on the number of duplicates available to IDE within the various groups at a given time.

5. *Cataloguing on slips and numbering of books and slips*

Cataloguing of the duplicates takes place only for such duplicates as are offered on exchange.

As distinct from the practice in the British Library the title pages of the books are not marked by a qualified cataloguer beforehand; the catalogue slips are typed directly from the books by the clerical staff or trainees. Books offering special problems are put aside for consultation with a qualified staff member.

In order to give a clear bibliographical description of the publications to make them readily identifiable, the catalogue slips are provided with the following information:

1. Author's surname and first name or initials
2. Title and subtitle
3. Publisher if a research institution
4. Doctoral dissertations are marked „diss." followed by the name and locality of the university
5. Place of publication
6. Year of publication
7. Title and number of series if any

Each catalogue slip is numbered, and a yellow slip carrying the same number is inserted in the book.

Nearly all duplicates of the IDE are printed in foreign languages. Lists of duplicates (books) in the field of the humanities may contain as much as 33 per cent titles in the Slavonic languages and in Finnish, a fact which naturally slows up the cataloguing procedure to some extent.

On one occasion, at the preparation of a list of dissertations printed in Danish and English, it proved possible to avoid cataloguing on slips (stage 5). The stencils were typed directly from the books. Naturally such a procedure is only possible if the material is uniform and printed in languages familiar to the cataloguer, in this case a trainee librarian.

6. *Alphabetical arrangement of slips and books*

After the cataloguing, the catalogue slips are arranged alphabetically, and the books are shelved in alphabetical order by means of the numbers on the catalogue slips and on the corresponding yellow slips in the books. This is a recently introduced system which facilitates and shortens the procedure of alphabetical shelving very considerably.

During the cataloguing phase it is a relief, and therefore a time-saving factor, that the order of the books is of no importance. Thus books may be conveniently put aside for consultation, and it is also possible to include incoming books in the collection which is being catalogued.

In order to avoid the catalogued books being displaced or taken away by visitors from Danish research libraries, who have free access to the rest of the collections, or by staff members needing a book for fulfilling a request, the catalogued books on the shelves are guarded by string until distribution.

7. *Proof-reading of slips*

Either before or after slips and books have been arranged alphabetically, the catalogue slips are proof-read by a qualified member of the staff.

8. *Typing of lists*

Typing out lists on stencils from the slips is done at the IDE by the clerical staff, whereas the printing of the lists takes place at the Royal Library (the National Library).

Lists have recently begun to be duplicated by a photocopying process rendering the use of stencils unnecessary, thus reducing the time spent on typing by approx. 25 per cent. The figures are those of the less time-consuming system.

All lists are provided with a number and an expiry date, and each page carries the number of the list.

9. *Selection of recipients for duplicates lists*

In 1975 IDE sent out a questionnaire to 325 foreign exchange partners, asking them to indicate their fields of interest. As 90 per cent returned the questionnaire, the IDE now has a fairly comprehensive survey of the interests of its foreign partners.

Traditionally, the Danish research library system is characterized by a high degree of specialization. Besides their general areas of interest several libraries are responsible for the acquisition of material within specialized narrow subject fields, either according to the Scandia Plan or by informal agreements.

The duplicates section has acquired a good additional knowledge of the needs and special interests of most Danish research libraries, which is kept up-to-date by personal contacts.

All exchange partners, domestic as well as foreign, are identified by code numbers; the addressees for each list may be selected by means of the addressing machine of the Institute.

Distribution

10. *Dispatch of lists to partners*

As already mentioned, the duplicates lists are sent to foreign and domestic partners simultaneously.

Lists for overseas addressees are sent by air mail as printed matter, in order to make it possible for these partners to observe the time limit of 3 months before the distribution of the publications requested from the lists takes place.

Owing to budgetary restrictions IDE now dispatches all lists as printed matter, since the postage for letters has been considerably raised. No decline in the number of requests has been observed, however.

11. *Receipt of requests*

Incoming requests are filed with the relevant list and brought forward on the date of distribution.

12. *Marking of requests on distribution list*

At the expiry of the time limit a distribution list for monographs is drawn up.

Letters or lists returned from exchange partners are numbered. These numbers are entered in the margin of a copy of the exchange list serving as a distribution list; every title of the list will show how many and which libraries require the title.

Use of the distribution list for allocation makes reference to the individual letters or lists returned by exchange partners superfluous, thereby simplifying the allocation procedure and saving time.

13. *Allocation of duplicates*

It is a principle at the allocation to give priority to the most important libraries. For foreign libraries the total amount of exchange material received by the IDE within all exchange activities (this applies first and foremost to duplicates and new periodicals) is taken into consideration.

Following the maxim „first come first served" would, naturally, be the easier and quicker way of distributing duplicates. However, this system is considered not just. Remote countries will be at a disadvantage; likewise large libraries are generally bound to be less privileged as their acquisition policy is complicated and the process time-consuming.

At the allocation of IDE's duplicates all exchange partners are allotted a share as far as possible. This means in practice that libraries ordering one or two titles are accommodated first. As for the remaining requests the priority ranking is applied.

The material is sorted out into piles, provided with address labels and form letters indicating that the books were forwarded in response to orders from the specified exchange list and the amount of duplicates dispatched. Any partners receiving no allotments to comply with their requests are notified to that effect.

The remaining duplicates (at an average 10 per cent of the titles) for which no orders were received, is kept for a couple of months in order to give exchange partners who have not been able to keep the time limit an extra chance of having a few of their requests fulfilled.

Some of the publications left over are offered a second time, part of them may be sold, and the rest is pulped. When new duplicates are received, the IDE tries to keep an eye on items which were ordered in vain from previous duplicates lists because too many libraries asked for the same title. Normally monographs offered are only available to the IDE in one copy.

14. *Keeping statistics*

Statistics are kept of the number of books and issues of periodicals forwarded by way of duplicates exchange to each partner. The figures are partly intended for the annual report of the Institute and partly serve to ensure a reasonable balance in the exchanges.

Still, the IDE in its exchange of duplicates is rather broadminded, refusing to follow a strict volume for volume policy.

15. *Packing and posting of duplicates*

Finally, the various consignments of duplicates are taken to the dispatch department of the IDE from where they are sent out to the addressees.

To most libraries in Copenhagen the duplicates are taken by IDE's driver. Some of the duplicates are sent in collective consignments to similar exchange centres abroad; in case no such centre exists, they are sent direct to the exchange partners.

In order to save administrative work and on the theory that the amounts for incoming and outgoing material will be approximately equal, IDE does not charge postage or freight, neither does it pay any freight expenses for incoming material.

It may be added that for the same time-saving purpose the amount of correspondence is kept to a minimum by using form letters, whenever possible.

Other ways of using duplicates

It is obvious that with its limited personnel the IDE is unable to prepare lists of 1,000 running metres a year. Generally 20 lists are issued a year. In the case of monographs each list covers an average of 600 titles. However, IDE has other ways of making use of its duplicates:

Visits from Danish research libraries are encouraged, and visitors are given free access to the shelves except for the collections which are catalogued for exchange lists. The visitors select the duplicates they need for their libraries.

In this way the IDE has contributed towards the building up of the stocks of recently established Danish university libraries and institutions and of Danish collections abroad.

On an average the IDE receives visitors from 30 institutions a year.

From time to time libraries of the developing countries or libraries which have suffered special privations apply to Denmark for help. Their applications are often addressed to the Ministry of Education, the Ministry of Cultural Affairs or to the National Library. In many cases the Exchange Institute is asked to contribute publications to meet the need of such applicants. A certain coordination of help for developing countries takes place.

The IDE has obtained permission to sell duplicates which are considered difficult or impossible to place by exchange. Instead of receiving money from antiquarian booksellers, the IDE will establish an account with them on which to draw for the benefit of Danish libraries which have delivered duplicates to IDE.

Occasionally IDE sends ,,special offers" of e.g. particularly valuable works or long runs of periodicals to Danish institutions or to such foreign partners as have provided the IDE with a large amount of valuable exchange material. Special offers are sent to one library at a time, which means that the recipient can be sure of receiving all the publications offered, unlike the duplicates lists.

From time to time the Institute is able to react to desiderata lists received from domestic and foreign libraries whose requirements it attempts to meet to the greatest possible extent.

Quite often domestic libraries apply to IDE by telephone asking for a specific publication, in which case IDE endeavours to fill the gap.

Conclusion

At the planning of the three papers on the economy of duplicates exchange it was agreed to analyse only the cost and/or time of the various work phases involved. Other factors such as storage, utilities, office expenses, etc. are included in the one of the other studies, however.

As for the benefit of duplicates exchange the aspect of re-use should not be the sole consideration. Keeping a statistical balance of duplicates dispatched and received is the only possible system, but it does not give the whole picture.

The intrinsic value of the duplicates is difficult to estimate. The monetary value is incalculable e.g. of issues of periodicals filling a gap in a run, or a book urgently needed by a library but out of print.

Duplicates exchange offers libraries unique (and cheap) opportunities of acquiring publications otherwise unobtainable.

	WORK PHASE	NO. OF ITEMS (BOOKS)	TIME PER ITEM (MINUTES)	NO. OF PERSONS	CATEGORIES* OF STAFF
1	Unpack				
2	Discard unwanted items	3650	1	2	A, Al
3	Sort and shelve received material in subject groups	2500	2	4	A, Al, B
4	Select duplicates for exchange lists		no calcu-lable time		
5	Catalogue books on slips and provide the books and slips with numbers	1000	2,3	2	B
6	Arrange slips and books alphabetically	2630	1,5	3	B
7	Proof-read slips	1200	0,3	3	A

* A: Staff members: librarians and head of duplicates section
Al: Head of dispatch section
B: Clerical staff and trainees

	WORK PHASE	NO. OF ITEMS (BOOKS)	TIME PER ITEM (MINUTES	NO. OF PERSONS	CATEGORIES* OF STAFF
8	Type list from slips [Printing of stencils]	2300	0,5	3	B
9	Select recipients for list				
10	Despatch lists to partners	1788	0,4	2	A, Al
11	Collect requests received		no calculable time		
12	Mark requests on distribution list	2488	0,3	3	A, B
13	Allocate duplicates	2488	0,6	3	A
14	Keep statistics of outgoing duplicates	1850	(0,0016)	1	A
15	Pack duplicates and post them to partners	2772	0,5	2	Al, B
			total: 9,4		

References

* Chief, I. D. E., Copenhagen

1 A Costing Analysis of the Gift and Exchange Section of the British Library Lending Division. By E. Tatterton, K. King and A. Allardyce based on an internal report by R. Steemson. Working Paper, IFLA, General Council, Štrbské Pleso, 1978. 14 p. See p. 75–95.

2 H. Rovelstad: The Economics of the Universal Serials and Book Exchange (USBE). Working Paper, IFLA, General Council, Štrbské Pleso, 1978. 15 p. See p. 109–117.

The Economics of the Universal Serials and Book Exchange (USBE)

Howard Rovelstad*

The Universal Serials and Book Exchange, Incorporated (USBE), is now, as it has been from its founding in 1948, a free-standing, self-supporting, non-profit agency responsible to a cooperative library membership for one principal program: the receipt and handling of surplus and duplicate publications so that they are available to the maximum degree possible to libraries of all kinds, on all levels of size and development, and in all countries of the world. To understand the economics or finances of USBE it is necessary to understand the management of USBE affairs with reference to sources of income, expenditures, and development of resources.

First, it is important to understand what USBE is not. USBE is not an agency of the United States or any other government; it is not an agency of a national library as are such programs in many other countries; it is not a profit-making, commercial venture.

1. *General Operating Structure*

As an independent self-supporting library service organization USBE operates a clearinghouse for the receipt, organization, and redistribution of surplus and duplicate publications, largely periodicals and serials but also including books, government documents, and other library materials. Members in the USBE complex include 1,763 member libraries (1,462 in the United States and Canada and 301 in other countries). A broad variety of kinds and sizes of libraries characterizes USBE: there are 940 special libraries, 751 principal libraries of academic institutions, and 72 public, state, and national libraries. Within the special library group, 243 are within academic institutions. Of the overall total, 60 % are supported by national or local governments, and 40 % by private organizations of which about half are commercial and half non-profit.

Members deposit materials in a central warehouse, select materials from the refined and organized clearinghouse collections for their own needs, and pay an annual membership fee, plus shipping charges and a handling fee for each publication they acquire. The requirement of „exchange" is based on the principle that a library should send to USBE any publications it is able to send, after meeting its own community needs, and that it should take from USBE anything it needs for which it is willing and able to pay at-cost handling fees; no numerical quid pro quo is required.

USBE processing of materials received is based on providing the best possible access for libraries with the least cost within USBE operations. Included in this standard are maintenance of the absolute minimum of records (e.g. alphabetical location taking the place of periodical cataloguing), flat handling fees based not on market value but on the type of publication involved (eliminating the necessity for costly pricing and accounting), and a requirement that all requests be firm orders so that the total operating costs can avoid the loss of returned shipments. Distribution methods are based on the fairest and most open methods possible, avoiding various burdens existing in some exchange distribution plans.

2. *Governance*

The governance of USBE is carried out by officers and the Board of Directors composed of the Executive Director of USBE and eleven members who are elected by member libraries and sponsoring library associations, each of which has one vote in the annual election. Officers and Board are elected from a slate of two candidates for each office, selected at large by a nominating committee and subject to additional nominations from the electorate.

At two or more meetings a year the Board votes on such matters of fundamental policy as the annual budget, recommendations on changes in the by-laws, changes in the fee schedule, and appointment of the chief executive officer. It makes recommendations to the staff on procedures and programs, and reviews current affairs of the organization, hears reports of the staff, committees and special studies.

3. *Operations*

Operations are carried out in two warehouse buildings in northeast Washington, D. C., with a total of 33,000 square feet of rented space and twelve miles of shelving. The staff of 55 consists of five administrative officers, ten supervisors, and 40 others in periodicals and books divisions, office, and receiving and shipping.

Stocks available for distribution consist of over four million publication items on hand at any one time, including 100,000 books, government documents, and other library materials such as microfilm and 4,000,000 periodical issues in 40,000 titles. About a million and a half items are received annually. Analysis of stocks shows these proportions:

Published in the United States 60 %, in other countries 40 %; 70 % in English, 30 % in other languages.

Subjects: health sciences 25 %, other science and technology 30 %, social science 25 %, arts and humanities 15 %, general and popular 5 %.

Date range: from beginning to the current for periodicals, with approximately 5 % in issues of the last six months; in-print books largely in the last fifteen years, wider range for out of print books and documents.

USBE's selection guide, which governs the titles and types of material requested from lists of offers and selected incoming materials, is a historical tally of materials requested by members. This means that for periodicals the selection process can be mechanical. Titles which the records show have been requested, either from USBE lists or directly, are retained in numbers appropriate to the number of requests; those not requested, or issues received in numbers in excess of the maximum needed are discarded and recycled. Over the period from 1948 through 1979, 47,100,000 publications were processed. Libraries requested and received 13,707,238 publications; of the remainder not shelved for future requests, 19,000,000 were discarded on receipt and 12,500,000 were discarded from inventory. New titles are generally kept for a test period. With both books and periodicals, materials on the shelves are constantly and progressively checked against the historical request record, and those no longer in demand are discarded. By applying these methods the publications bank is kept constantly as current as possible and under a control of total size.

4. *Maximum Accessibility*

The processes developed by USBE form an integral part of the resource represented by the organization in that these methods have given libraries maximum accessibility to available materials and the best possible return for each dollar they spend.

Member libraries looking to this collection for resources they need may send want lists or purchase orders in any form convenient to them to USBE, or check off needed items on the lists they receive each month from USBE, or make personal selections from USBE shelves. The most valuable contribution to libraries continues to be serving as the source for issues of periodicals or serials to fill gaps and to replace damaged or lost issues. Also, new institutions and institutions with new programs that require building up of new collections find it possible to acquire files of journals, either from the beginning up to the current year, or perhaps for the last five or ten years through current issues. Both new and established institutions use the monthly lists of books, government documents, indexes, and other publications to add new items and replace old ones, as well as to acquire out of print materials. In the United States, because of the growing incidence of fees charged by libraries for interlibrary loan, many libraries use the regular USBE single-issue acquisition service as a substitute for interlibrary loan. During a test period libraries ordering items to use in this way were successful in 60% of their orders.

5. *Statistical and Cost Studies*

The framework of the USBE statistics is that of a self-contained operation receiving publications from its member libraries in a central warehouse, reviewing and selecting from the incoming publications those on request or to be shelved to await future requests, shelving and listing the selected items, providing items from the shelves on request, and shipping them out. Member libraries pay all the costs of the operation by means of a $ 25 per year membership fee and a standard handling fee for each publication they request and receive.

As this paper is written, cost and item studies have been under way for three years:

1. To provide general information to the library on USBE's basis for handling fees and membership charges.
2. To provide guidance for fee changes.
3. To provide data useful to increase efficiency and reduction of costs wherever possible, as inflation continues and library budgets tighten.
4. To prepare USBE for future automation to allow the development of better service to libraries equipped with machine readable records, and linkage with library networks both domestic and international.
5. To prepare data on which to base decisions as to modification of any of USBE's traditional functions whose cost may outweigh their value as services to libraries.
6. To give USBE a basis for planning for future developments including added services.
7. To provide information which may be useful for librarians anywhere in the world planning comparable network operations or studying the economics of exchange and of other systems serving libraries in acquisitions and interlibrary loan.

2. The following statistics show the time and cost per item of providing publications on demand from libraries from an analysis prepared from the operations records of the calendar year of 1977.

a. Man hours involved in 1977 operations

	Total Man Hours	Hours Per Item Distributed
Administration and Overhead	27,004	.080
Receiving	5,859	.017
Reviewing, Shelving, Collection Maintenance	25,731	.076
Processing of Requests	39,075	.116
Shipping	8,673	.026
	106,363	.315 (18.9 minutes)

b. Dollar costs of USBE operations in 1977

	Salaries	Salary-Related Costs*	Housing & Equip-ment	Publicity & Board Expenses	Other Costs	Total	Cost Per Item Distri-buted
1) Administration & Overhead	$ 188,358	$ 22,203				$ 210,561	$ .62
			$ 80,510			80,510	.23
				$ 23.225		23,225	.07
							.41
					$ 137,709	137,709	
2) Receiving	25,394	3,265				28,659	.09
3) Reviewing, Shelving, Collection Maintenance	155,415	13,714				129,129	.38
4) Processing of Requests	162,772	19,591				182,363	.54
5) Shipping out	54,631	6,530				61,161	.18
Totals	$ 546,570	$ 65,303	$ 80,510	$ 23,225	$ 137,709	$ 850,317	$ 2.52

* Payroll taxes, retirement expense, health insurance, etc.

In this breakdown only salary and salary-related costs are given for the various separate functions; all other costs are listed in Overhead.

3. One of USBE's chief services to libraries lies in the fact that it accepts and sorts through mixed and miscellaneous material in order to retrieve and make available to libraries that portion of the whole which is useful for their collections.

   Pre-sorting is now performed by many libraries; this process may be increased in the future. If USBE inventory controls are automated the cost of handling unwanted materials may be substantially reduced. In the following preliminary figures, salary costs include both salaries and salary-related costs (which form part of Item 2) and 3) in 2b above) for receiving and collection maintenance.

Publications (Numbers are annual counts)		Receiving Cost	Collection Maintenance Cost	Total	Per Item Cost
Discarded on receipt	100,000	$ 2,293	$ 6,220	$ 8,513	$ .09
Discarded after checking shelves	120,000	2,866	7,747	10,613	.09
Discarded from shelves as surplus	200,000	4,585	36,000	40,585	.20
Supplied to libraries	330,000	7,738	32,102	39,840	.12
Shelved but not yet distributed	480,000	11,177	47,060	58,237	.12
		$ 28,659	$ 129,129		

4. Studies of USBE costs are proceeding and have already resulted in some changes as well as projected modifications and identification of areas for further study.

   a. Types of orders had been previously charged on a scale acknowledging (although not precisely accounting for) the fact that a single issue is more expensive to produce than one issue which is part of a larger order; USBE fees at the beginning of 1979 were $ 3.50 each for a single issue (also for recent issues and first issue in a multiple order) and $ 2.25 each for multiple issues. Determining that maintaining the varying charges was too expensive both for USBE and for the ordering libraries, the staff introduced an across-the-board fee of $ 2.50 per issue (increased to $ 2.75 in November).

b. It is apparent that holding and maintenance of publications not frequently requested affects the per-item cost of those distributed. USBE is therefore reducing the numbers and character of the lower-demand titles in order to cut costs; the process of identification of the higher- and lower-demand titles will increase the possibility of further flexibility and economy in the future.

c. Studies leading to at least partial automation of records are resulting in preliminary procedural improvements; these and automation itself will not decrease costs but will make a larger distribution both possible and attractive, thus leading to reductions in the per-item fee, as is illustrated below:

	Item Count	Total Cost	Per Item Cost (average of all publication fees)
1979 Item Distr. (Est.)	330.000	$ 960,000	$ 2.91
Speculated Additional distr.	60,000	Total cost with automation	
	390,000	$ 1,100.000	$ 2.82

The original purpose in any library's use of USBE is still the principal one: to acquire publications needed to build collections, fill gaps, and replace missing items. Over the years libraries have developed other uses for USBE service: for example, as a source for publications wanted by individual staff or faculty members, or for publications the library wants to use in filling needs of its exchange or interlibrary loan partners. USBE distribution facilities are being considered in the planning now under way for a U.S. National Periodicals System. The original purpose of the world-wide library community has been carried out by USBE in its retrieval for library use of more than 13,000,000 publications, which would otherwise have been lost. The USBE staff and the libraries it serves continue to search for more economical and more numerous ways it may serve its members.

* Director of Libraries Emeritus, University of Maryland, College Park, Maryland, USA

Rising Book Prices and the Exchange of Publications

Maria Razumovsky*

Rising prices of books and periodicals are a heavy burden to libraries throughout the world. A comparison of book prices in the last 10 years has shown that in some countries and in some book categories the increase in prices has amounted to 50 %, 100 % and sometimes even to 250 %. We have to face the fact that the cost of book and non-book material will continue to rise steadily or even steeply and that very often it will exceed increases in acquisition budgets. Since exchange is one of the ways to acquire foreign material, exchange services and the exchange sections of libraries are directly confronted with this problem.

In order to find out to what extent increases in book prices influence the exchange policy of libraries, a questionnaire, drafted at the IFLA Congress in Brussels and considerably improved by the comments of SC members, especially by Mr. Allardyce, was sent out to national libraries throughout the world and to a selection of university and large research libraries and exchange centres, in total to 84 addresses. 59, i.e. nearly 70 %, took the trouble to study this questionnaire thoroughly and to reply to the questions. I would like to thank all of them sincerely for their efforts and their contribution to our study. The survey was closed on 1 July 1978.

The replying libraries can be classified as follows:

National and large research libraries	30
University libraries, some of them acting also as national libraries	21
Special libraries and collections	5
Exchange centres	3

Of the 59 replies, 15 come from East Europe, 32 from West Europe and only 12 from libraries outside Europe. It is to be regretted that both the United States and Canada are completely missing.

1. *Is exchange a regular part of the acquisition policy of your library?*

To this question, only 2 said "no": The Schweizerische Landesbibliothek, Berne, which receives by exchange only bibliographies, and the National Library of Ireland, Dublin, that exchanges only official publications. For the other libraries, exchange is a regular part of their acquisition policy.

2. *Does your library keep exchange statistics?*

Yes: 48
No: 9

3. *Which of the following categories of material do you regularly obtain by exchange with other libraries and institutions abroad?*

Recent monographs:	32
Current periodicals:	52
Official publications:	35
Dissertations:	31
Secondhand and duplicate materials:	28
Non-book trade materials:	14
Other items:	11

No library declared that no material at all is being received by exchange.

4. *What % of your total intake comes by exchange?*

a) Periodical titles:
0,3 % : Florence
1 – 5 %: 6 (Munich, Berne, Baghdad, Aarhus, Lund, Pretoria)
5 – 15 %: 13
15 – 25 %: 9
25 – 50 %: 16
more than 50 %: 8

b) Other items:
0 % : 3 (Berne, Dublin, Florence)
1 – 5 %: 5
5 – 15 %: 17
15 – 25 %: 8
25 – 50 %: 10
more than 50 %: 8

Among those which receive more than 50 % of their intake by exchange are the Academy of Sciences Library Leningrad (90 %), the Lenin State Library Moscow (80 %), the University Library Uppsala (70 %), the Saltykov– Ščedrin Library Leningrad (67 %), the National Library Baghdad (60 %) and specialized collections: Institute for Cultural Relations Finland-Soviet Union Helsinki (68.2 %), British Library, Oriental Department (64.5 %).

5. *With how many institutions abroad do you have exchange agreements?*

It is no surprise that, with the exception of Canberra (about 6000 agreements), it is mostly libraries from East Europe that have the greatest number of exchange agreements: Academy of Sciences Library Bucharest (about 10.000), libraries in the Soviet Union (about 3000). In the next group (1000–2000 agreements) there can be found Oslo (1450), Zürich (1280), the Deutsche Staatsbibliothek Berlin (1230), Universitätsbibliothek Leipzig (1202), Bibliothèque Nationale Paris (1100), the Bodleian Library, Oxford, the University libraries of Gothenburg and Liège. The greatest number of libraries (14) have between 200 and 500 exchange agreements and 7 have less than 100.

6. *Is it a regular practice of your library to purchase new items published in your country for exchange with libraries abroad?*

From the replies it turns out that not less than 40 libraries as well as the exchange centres purchase new items for their exchange partners and that only 16 do not. Some libraries purchase with restrictions ("Only dissertations"; "Only for exchange with East Europe").

7. *Approximately what percentage of your total acquisitions budget (including any sums allocated separately for exchange purpose) is spent on the purchase of this material?*

Nothing:	3
Less than 5 %	19
5 - 10 %	8
10 -20 %	10
More than 20 %	6
Not known, no reply	10

To those that spend less than 5 % belong most of the West European libraries, Tokyo and Canberra; among those that spend more than 20% are the libraries in the Soviet Union and most of the specialized collections.

8. *Is the percentage you gave in your answer to question 7 likely to change in the near future?*

Increase	19
Remain fairly constant	27
Decrease	5
No reply	8

This is a very positive outlook, especially as the 5 libraries that think that their exchange budget will decrease, are those which are already spending a very high percentage for exchange.

9. *If you have been forced to reduce the general level of acquisitions, through economic stringency, have the effects had a greater or lesser impact on purchasing for exchange than on other purchasing?*

Greater impact	3
About the same	15
Lesser impact on exchange	4
Have not yet had to reduce the level of acquisitions	29

That 57 % of these libraries have not yet been forced to reduce the level of their acquisitions is interesting and might be thought to be encouraging.

10. *Have you been asked by your exchange partners abroad for material published in your country which you have been unable to send*

a) for financial reasons? b) for other reasons?

52 libraries gave an affirmative answer to this question, of whom 20 stated that it was for financial reasons. Other reasons were that the requested material was out of print (quoted most often), that publications not for sale were requested, that the material was censored and that the publications offered by exchange were of no interest to the library.

11. *If you are unable to comply with all orders of your exchange partners, how do you face the problem?*

By reducing the number of partners:	5
By reducing the material sent on exchange:	11
By excluding certain types of material:	12
By offering the material you have in your stock of duplicates instead of the material wanted:	20
We have not yet been in such a position:	1

12. *Do you think that as a consequence of rising book prices the exchanges of your library will become in future*

More important:	20
Remain fairly constant:	29
Less important:	2
No reply:	8

As we see, the replying institutions have on the whole a positive opinion on the future of exchange. They know that this way of acquisition is an important help, although the importance is different in different parts of the world. But there were also some critical remarks, which seem to be worth taking into consideration:

- We exchange mostly serial publications and only when those offered by exchange partners are really of use to us. Each offer is treated on its merit. We then offer a list of what we have available . .. (Pretoria)
- With some very important exceptions, we find exchange a less efficient mode of acquisition than purchase. (London)
- We are intentionally reducing our exchange these days, since we find that, expecially with severe financial restrictions, purchasing is the more practical way of getting the material. You save the costsof storing and handling and distributing the exchange material, and you are sure to get just what you need and nothing else. (Lund)
- Exchange becomes more expensive not only because of the continuous increase of prices, but also because of the rise of binding and shelving costs, especially for periodicals and newspapers. Therefore not the expansion of exchange should be propagated, but reduction to its optimal scope. Exchange may tempt a library to send or order less or not needed literature, just for the sake of a settled exchange balance. (Wien)
- Acquisition by exchange is the most labour intensive and cost intensive mode of acquisition. (Leipzig)

* Director, Acquisitions Department, Österreichische Nationalbibliothek, Vienna, Austria

Comment on the paper by Razumovsky

Peter Genzel

This study is very important; the survey explored one of the crucial problems of today's acquisition and exchange policy.

1. The result is most interesting and encouraging. In spite of rising prices a positive outlook prevails:
   - Question 9: 29 libraries need not reduce the level of their acquisitions, and for another 15 the purchase of literature to be exchanged remains about the same = 44 altogether out of 59 replies to the questionnaire.
   - Question 8: 19 libraries will spend even more on the purchase of exchange material than before, and in 27 cases the level will remain fairly constant = 46 altogether.
   - Question 12 (which reflects question 8): For 20 libraries exchanges will become more important, and in 29 cases it will at least remain fairly constant = 49 altogether.

   The total figures are nearly all the same in all 3 instances.
2. Another result, not indeed surprising, but not explicitly pointed out, can be read between the lines: There is an inevitable contrast of libraries in East with those in West Europe caused by the money question:
   - The prices of new books and periodicals are rising almost exclusively in Western countries whereas they remain fairly constant in East Europe. To some extent this affects also official publications and other material.
   - Budgets for purchase through the book trade are occasionally limited.
   - Therefore many libraries in East Europe are interested in extending their exchange to book trade publications, beyond the traditional exchange of non-commercial publications (official publications, duplicates, etc.). For this they have sufficient local currency and staff, and they have a good number of important publications to offer.
   - If an equivalence is aimed at with regard to the exchange of commercial publications, the question of money comes into play again. Because of the different price levels priced exchange is disadvantageous to East European libraries.

   The whole problem runs like a red thread through the study:
   - The fact that 15 replies out of 59 came from East Europe (p. 118, 29th line), proves their vital interest.
   - Many of them receive more than 50 % of their intake by way of exchange (question 4).
   - They have the greatest numbers of exchange partners (question 5).

- Some West European libraries even when purchase for exchange is restricted do buy literature for exchange with East Europe (question 6).
- Many East European libraries spend more than 20 % of their budget on the purchase of exchange material, in contrast to many West European libraries with less than 5 % (question 7).

3. This last statement illuminates another basic question mentioned in the critical remarks (p. 122). Of course exchange is a labour and cost intensive mode of acquisition, and purchase is a more practical way. However, as stated above, many libraries are dependent on the exchange of commercial publications, and they must find ways and means for this.

   In an effort to cushion the impact of rising binding and shelving costs, especially for periodicals, some libraries see the need to reduce their exchanges. But periodicals must be bound in any case, whether purchased or received through exchange. What they have in mind is to prevent receiving unwanted material in exchange. But this can be avoided if the principles of acquisition are strictly applied and precise agreements are made before an exchange is started.

   I am conscious of the fact that there are even more sides of this problem, but the above-mentioned questions are the central points.

Series IFLA Publications
Edited by Willem R. H. Koops

---

3 **Le contrôle bibliographique universel dans les pays en développement.** Table ronde sur le contrôle bibliographique universel dans les pays en développement, Grenoble, 22–25 août 1973. Ed. par Marie-Louise Bossuat, Geneviève Feuillebois, Monique Pelletier. 1975. 165 p. DM 38.00, IFLA members DM 29.00. ISBN 3-7940-4423-1

4 **National and International Library Planning.** Key papers presented at the 40th Session of the IFLA General Council, Washington, DC, 1974. Ed. by Robert Vosper and Leone I. Newkirk. 1976. 162 p. DM 38.00, IFLA members DM 29.00. ISBN 3-7940-4424-X

5 **Reading in a Changing World.** Papers presented at the 38th Session of the IFLA General Council, Budapest, 1972. Ed. by Foster E. Mohrhardt. 1976. 134 p. DM 28.00, IFLA members DM 21.00. ISBN 3-7940-4425-8

6 **The Organization of the Library Profession.** A Symposium based on contributions to the 37th Session of the IFLA General Council, Liverpool 1971. Ed. by A.H.Chaplin. 1976. 2nd edition. 132 p. DM 28.00, IFLA members DM 21.00. ISBN 3-7940-4300-X

7 **World Directory of Administrative Libraries.** A guide of libraries serving national, state, provincial, and Länder-bodies, prepared for the Sub-section of Administrative Libraries. Ed. by Otto Simmler. 1976. 474 p. DM 60.00, IFLA members DM 45.00. ISBN 3-7940-4427-4

8 **World Directory of Map Collections.** Compiled by the Geography and Map Libraries Sub-Sections. Ed. by Walter W. Ristow. 1976. 326 p. DM 48.00, IFLA members DM 36.00. ISBN 3-7940-4428-2

9 **Standards for Public Libraries.** 1977. 2nd corrected edition. 53 p. DM 16.80, IFLA members DM 12.60. ISBN 3-7940-4429-0

10 **IFLA's First Fifty Years.** Achievement and challenge in international librarianship. Ed. by Willem R.H. Koops and Joachim Wieder. 1977. 158 p. DM 36.00, IFLA members DM 27.00. ISBN 3-7940-4430-4

11 **The International Federation of Library Associations and Institutions.** A Selected List of References. 2nd edition, revised and expanded. Comp. by Edward P. Cambio. 1977. VI, 52 p. DM 16.80, IFLA members DM 12.60. ISBN 3-7940-4431-2

12 **Library Service to Children: An International Survey.** Edited for the Section of Children's Libraries by Colin Ray. 1978. 158 p. DM 36.00, IFLA members DM 27.00. ISBN 3-7940-4432-0

13 **Allardyce, Alex: Letters for the International Exchange of Publications.** A Guide to their Composition in English, French, German, Russian and Spanish. Ed. by Peter Genzel. 1978. 148 p. DM 36.00, IFLA members DM 27.00. ISBN 3-7940-4433-9

14 **Resource Sharing of Libraries in Developing Countries.** Proceedings of the 1977 IFLA/Unesco Pre-Session Seminar for Libraries from Developing Countries. Antwerp University, August 30–September 4, 1977. Ed. by H. D. L. Vervliet. 1979. 285 p. DM 36.00, IFLA members DM 27.00. ISBN 3-598-20375-6

15 **Libraries for All / Bibliothèques pour tous.** A World of Books and Their Readers / Le monde du livre et de ses lecteurs. Papers presented at the IFLA 50th Anniversary World Congress, Brussels 1977. Ed. by Robert Vosper/Willem R. H. Koops. 1980. 163 p. DM 36.00, IFLA members DM 27.00. ISBN 3-598-20376-4

16 **Library Service for the Blind and the Physically Handicapped: An International Approach.** Key Papers presented at the IFLA Conference 1978, Strbske Pleso, CSSR. Ed. by Frank Kurt Cylke. 1979. 106 p. DM 30.00, IFLA members DM 22.50. ISBN 3-598-20377-2

17 **Guide the Availability of Theses.** Ed. by D. H. Borchardt and J. D. Thawley. 1981. 443 p. DM 68.00, IFLA members DM 51.00. ISBN 3-598-20378-0

18 **Studies in the International Exchange of Publication.** Ed. by P. Genzel. 1981. 125 p. DM 32.–, IFLA members DM 24.–. ISBN 3-598-20379-9

19 **Public Library Policy.** Proceedings of the IFLA/UNESCO Pre-Session Seminar Lund, Sweden August 20–24, 1979. Ed. by K.C. Harrison. 1981. 152 p. DM 36.–, IFLA members DM 27.–. ISBN 3-598-20380-2. In preparation.

K · G · SAUR München · New York · London · Paris
POB 71 10 09 – D-8000 München 71 – Tel. (089) 79 89 01 – Telex 5212067 saur d

International Federation of Library Associations and Institutions

IFLA Annuals
Proceedings of the General Council Meetings. Annual Reports

IFLA Annual 1980 Proceedings of the 46th Council Meeting, Manila, 1980. Annual Reports 1981. 219 p. ISBN 3-598-20661-5. DM 58.–, IFLA members DM 28.50

IFLA Annual 1979 Proceedings of the 45th Council Meeting, Copenhagen, 1979. Annual Reports by W. Koops/W. Coops. 1980. 232 p. ISBN 3-598-20660-7. DM 58.00

IFLA Annual 1978 (44th Meeting, Strbske Pleso, 1978) 1979. 197 pages. ISBN 3-598-20659-3. DM 48.00

IFLA Annual 1977 (43d Meeting, Brussels: Libraries for All: one World of Information, Culture and Learning). 1978. 276 pages. ISBN 3-7940-4303-0. DM 58.00

IFLA Annual 1976 (42nd Meeting, Lausanne). 1977. 266 pages. ISBN 3-7940-4302-2. DM 58.00

IFLA Annual 1975 (41st Meeting, Oslo: The Future of International Library Cooperation). Ed. by W.R.H. Koops, P.Havard-Williams, W.E.S. Coops. 1976. 232 pages. ISBN 3-7940-4301-4. DM 58.00

IFLA Annual 1974 (40th Meeting, Washington: National and International Library Planning). Ed. by W.R.H. Koops, P.Havard-Williams, W.E.S. Coops, 1975. 314 pages. ISBN 3-7940-4300-6. DM 58.00

IFLA Annual 1973 (39th Meeting, Grenoble: Universal Bibliographic Control). Ed. by W.R.H. Koops, P.Havard-Williams, W.E.S. Coops. 1974. 256 pages. ISBN 3-7940-4299-9. DM 54.00

IFLA Annual 1972 (38th Meeting, Budapest: Reading in a Changing World). Ed. by W.R.H. Koops, P.Havard-Williams, W.E.S. Coops. 1973. 252 pages. ISBN 3-7940-4298-0. DM 48.00

IFLA Annual 1971 (37th Meeting, Liverpool: Organisation of the Library Profession). Ed. by P.Havard-Williams, W.R.H. Koops, H. J. Heaney. 1972. 239 pages. ISBN 3-7940-4297-2. DM 48.00

IFLA Annual 1970 (36th Meeting, Moscow: Libraries as a Force in Education). Ed. by Anthony Thompson. 1971. 336 pages. ISBN 3-7940-4296-4. DM 48.00

IFLA Annual 1969 (35th Meeting, Copenhagen: Library Education and Research in Librarianship). Ed. by Anthony Thompson and S. Randall. 1970. 289 pages. ISBN 3-7940-4295-6. DM 48.00

IFLA Directory 1979

1979. 197 pages, 1 folded leaf. ISBN 3-598-20779-4. DM 32.00

The Directory gives addresses with telephone and telex numbers of all IFLA offices, bodies, members, etc. It also contains statutes, conditions for membership and a list of publications.

Universal Bibliographic Control

A Long Term Policy – A Plan for Action

By Dorothy Anderson. 1974. 87 pages. ISBN 3-7940-4420-7. DM 16.80, IFLA members DM 12.80

This study was originally prepared as a working document to be presented by IFLA to the Unesco Intergovernmental Conference on the Planning of National Overall Documentation, Library and Archives Infrastructures.

The International Exchange of Publications

Proceedings of the European Conference held in Vienna, April 24–29, 1972

Edited by Maria Schiltman. 1973. 135 pages. ISBN 3-7940-4311-1. DM 28.00, IFLA members DM 21.00

K · G · SAUR **München · New York · London · Paris**
POB 71 10 09 – D-8000 München 71, Tel. (089) 79 89 01 – Telex 5212067 saur d

Bei Fragen zur Produktsicherheit wenden Sie sich bitte an:
If you have any questions regarding product safety,
please contact:

Walter de Gruyter GmbH
Genthiner Straße 13
10785 Berlin
productsafety@degruyterbrill.com